World Livestock 2011
Livestock in food security

FOOD AND AGRICULTURE ORGANIZATION OF
THE UNITED NATIONS

Rome, 2011

CONTRIBUTORS

Editor: A. McLeod

Copy editing: N. Hart

Design: C. Ciarlantini

The following people contributed to the content of this document: V. Ahuja, C. Brinkley, S. Gerosa, B. Henderson, N. Honhold, S. Jutzi, F. Kramer, H. Makkar, A. McLeod, H. Miers, E. Muehlhoff, C. Okali, C. Opio, I. Rosenthal, J. Slingenbergh, P. Starkey, H. Steinfeld, L. Tasciotti.

The following people provided ideas or references or contributed to the review process: D. Battaglia, J. Custot, K. de Balogh, N. de Haan, P. Gerber, D. Gustafson, I. Hoffmann, J. Lubroth, P. Kenmore, R. Laub, J. Otte, T. Raney, M. Smulders, P. Roeder.

RECOMMENDED CITATION

FAO. 2011. *World Livestock 2011 – Livestock in food security.* Rome, FAO.

COVER PHOTOGRAPHS:

©FAO/Giuseppe Bizzarri; ©FAO/Giulio Napolitano; ©FAO/Kai Wiedenhoefer; ©FAO/Noel Celis

ISBN 978-92-5-107013-0

Contents

List of Tables

List of Figures

List of Boxes

Foreword

Feeding the world's poor is one of the most pressing challenges of the present day, as human populations grow and put increasing strain on natural resources. Livestock have an important part to play, as they provide high-quality protein to consumers and regular income to producers. To fulfil their potential sustainably, they must be managed with care. Water, fossil fuel and grain are used in rearing animals, and there is an urgent need to use these resources more efficiently, to recycle and reduce waste, and to create a positive balance sheet in livestock's contribution to global food supplies.

One of the hardest challenges for food security is ensuring that all who need food have the means to buy it, particularly when volatile economies and natural disasters make already weak livelihoods even more unstable. Here livestock make a vital contribution as generators of cash flow and economic buffers, provided that market chains are organised to provide openings for small scale producers and traders and those in remote areas.

Livestock perform a variety of functions in different human societies. Communities dependent on livestock, those who practice mixed farming on a small scale, and consumers in cities, each have specific demands on farm animals and their products and distinct food security concerns. Different geographical regions also have their own perspectives, with the emerging economies acting as growth engines and the developed countries driving food safety and environmental regulations. All share a need for food systems to be sustainable and resilient. Each region and type of community will have an influence in shaping livestock's contribution to the food security of the future.

Samuel Jutzi
Director
Animal Production and Health Division
FAO

Acronym list

ACI	Agrifood Consulting
ABARE	Australian Bureau of Agricultural and Resource Economics
BFREPA	British Free Range Egg Producers Association
BMI	Body mass index
BSE	Bovine spongiform encephalopathy
CBPP	Contagious bovine pleuropneumonia
CFS	Committee on World Food Security
COMESA	Common Market for Eastern and Southern Africa
EC	European Commission
EU	European Union
FAO	Food and Agriculture Organization of the United Nations
FMD	Foot-and-mouth disease
FARM-Africa	Food and Agriculture Research Management-Africa
GDP	Gross domestic product
Hh	Household
HPAI	Highly pathogenic avian influenza
ICASEPS	Indonesian Center for Agro-socioecoomic and Policy Studies
IFAD	International Fund for Agricultural Development
IGAD	Intergovernmental Authority for Development (East Africa)
IMF	International Monetary Fund
Kcal	Kilocalorie
LEAD	Livestock, Environment and Development
LEGS	Livestock Emergency Guidelines and Standards
MDG	Millennium Development Goal
N	Nitrogen
NDDB	National Dairy Development Board (India)
OECD	Organisation for Economic Co-operation and Development
OIE	World Organisation for Animal Health
P	Phosphorus
PRSP	Poverty reduction strategy paper
RIGA	Rural income generating activities
RVF	Rift Valley fever
SFU	Sheep forage unit
SOFA	State of Food and Agriculture
TB	Tuberculosis
TLU	Tropical livestock unit
UNFPA	United Nations Population Fund
UNSC	UN Standing Committee on Nutrition
UNU	United Nations University
WHO	World Health Organization
WFP	World Food Programme

Overview

Although much has been said about livestock's role in achieving food security, in reality, the subject has been only partially addressed and no current document fully covers the topic. This report is an attempt to fill the gap.

It expands the 2009 State of Food and Agriculture (SOFA) (FAO, 2009b) section which examined the multiple roles played by livestock in the food security of the poor and advocated for support of smallholders, both in responding to opportunities in livestock production and in finding other opportunities within a broad rural development strategy.

Recognizing that food security is central to international development – and to the mandate of the Food and Agriculture Organization of the United Nation (FAO) – the report tells the story of livestock and food security from three perspectives.

LIVESTOCK AND GLOBAL FOOD SECURITY

The first section of the report presents a global overview, examining the role that livestock play in various dimensions of food security. It describes the place of livestock products in human nutrition, the contribution of livestock to the world food supply and its stability, and it discusses the way that livestock can affect food access, particularly for poor families, as a direct source of food and a source of income.

"**Measuring food security**" explains the ways that food security is defined and measured, and provides an overview of trends in food security worldwide and by region. It is intended as background for those not familiar with the subject.

"**Livestock food in the diet**" describes the place of livestock products in human nutrition. Noting that the definition of food security includes the need for an "adequate diet", it discusses the positive contributions that livestock

products can make to the diet and the possible consequences of over consumption.

"**Livestock and food supply**" describes the contribution of livestock to the world food supply, directly through production of meat, milk and eggs, and indirectly through supplying traction and manure to cropping. It examines the factors that cause food supplies to become unstable and ways that livestock can mitigate damaging effects. It also reviews the causes of instability in the supply of livestock source foods.

"**Access to food**" deals with food access, examining the way livestock can improve household and individual access to food, particularly for poor families, and provide food and income. It also reviews the impact of gender dynamics on access to livestock source foods within families and the ability of families to earn income from livestock.

THREE HUMAN POPULATIONS, THREE FOOD SECURITY SITUATIONS

The second section shifts from the global level to a human perspective, examining the way in which livestock contributes to the food security of three different human populations – livestock-dependent pastoralists and ranchers, small-scale mixed farmers and urban dwellers. The chapters briefly describe the main issues each population faces and then introduce related case studies to examine certain points in more practical detail.

"**Livestock dependent societies**" examines the situation faced by livestock-dependent societies, including pastoralists, who are the main focus of the chapter, and ranchers. In both systems, livestock provide the foundation of livelihoods and contribute to food security both locally and globally. Mongolia is used as a case study, because it has a long history of livestock dependence but, as in other livestock dependent societies, is changing in response to external pressures and new opportunities.

"Small-scale mixed farmers" focuses on small-scale mixed farmers for whom livestock provide both food and the means to obtain it. For these farmers, livestock are an important, but not necessarily the most important, part of their livelihood portfolios. The chapter reviews the contributions that livestock in these systems currently make to food security, the constraints to expanding their contribution and the prospects for small-scale mixed farming. Nepal is used as a case study because it has a large number of small-scale mixed farmers who face strong resource and market constraints and therefore provide a good illustration of the challenges they face in increasing production from their farms.

"City populations" considers the case of city populations, growing in number worldwide. For the inhabitants of large cities, animal products are essentially a commodity to be consumed – unlike the livestock-dependent and small-scale mixed farmers who are both producers and consumers. The chapter considers the place of livestock products in the urban diet, the logistics of feeding city populations, and the factors driving the livestock production systems and market chains that supply cities. It compares the approaches taken by different countries, drawing the most detailed information from China and the United States of America (USA), which define their "foodsheds" – the surrounding area that can provide food for a city – in two very different ways.

FEEDING THE FUTURE

The final part of the report looks to the future. It discusses the expected demand for livestock source food and the way that increased demand can be met with ever more limited resources. It reviews the drivers that led to the livestock revolution, how these have changed and what the implications will be for livestock contributing to food security.

"Producing enough food" considers the task of producing enough food for future populations. It reviews FAO's projections of growth in demand for livestock source foods between 2010 and 2050, discusses the assumptions that were made in these calculations and the implications of any changes. It argues that reducing various forms of waste in livestock food systems will be an essential component of meeting future demand. Returning to the three populations of the previous section, it reviews where the emphasis may lie for each one in reducing waste and increasing efficiency.

"Building resilience" looks at the possibilities for improving resilience in livestock food systems and the increasing concern about the instability of food supply and access during what are termed "protracted crises". Livestock food systems must be prepared to respond to these crises, which will require building an increased capacity to deal with change and recover from shocks. The chapter reviews some of the factors that may create vulnerability in livestock food systems and suggests ways to mitigate them.

"Conclusions" summarizes the main messages of the entire report. It concludes that livestock make a positive contribution to food security but, at the same time, suggests that livestock need to be managed carefully to avoid externalities.

Livestock and global food security

©FAO/Vasily Maximov

Measuring food security

In 1996, the World Food Summit Rome Declaration set a target to reduce hunger by half by 2015.

In 2000, the United Nations Millennium Summit re-affirmed the target, making the halving of extreme hunger and poverty its primary Millennium Development Goal (MDG).

Notwithstanding these optimistic goals, in 2010, 925 million of the world's inhabitants still suffered from chronic hunger, and world food security was an uncertain prospect. Predictions concerning future food security must factor in assumptions about the growth of the economy, the distribution of income, the possibility of dealing with environmental challenges, and the political and logistical capacity to make food accessible everywhere, to everyone.

SIX DIMENSIONS

FAO defines four "pillars" of food security and two temporal dimensions related to food insecurity, all of which must be addressed in efforts to reach hunger reduction targets. The four pillars, detailed in Box 1, include: **food availability** which refers to food supply, and **food access** which means the ability of people to obtain food when it is available. As both availability and access must be stable, the third pillar, **stability**, refers to ensuring adequate food at all times while the fourth, **utilization**, incorporates food safety and nutritional well being.

Pillars. Paying simultaneous attention to all four pillars is a constant challenge. Sufficient food can be produced today to feed everyone in the world, but it is not always available in every country, let alone every community. Some countries produce enough food to be self sufficient while others rely on imports, meaning that when international prices rise or global value chains break down, the food supply becomes unstable. Even when food is available, many people cannot afford to buy what they need for a healthy diet and, in parallel, prices that can be paid by the poorest consumers may not be sufficient to provide a living for producers. Waste in food chains from oversupply and spoilage adds to costs and reduces the amount available to eat.

BOX 1

FOUR PILLARS OF FOOD SECURITY

Food Availability: The availability of sufficient quantities of food of appropriate quality, supplied through domestic production or imports (including food aid).

Food access: Access by individuals to adequate resources (entitlements) for acquiring appropriate foods for a nutritious diet. Entitlements are defined as the set of all commodity bundles over which a person can establish command given the legal, political, economic and social arrangements of the community in which they live (including traditional rights such as access to common resources).

Stability: To be food secure, a population, household or individual must have access to adequate food at all times. They should not risk losing access to food as a consequence of sudden shocks (e.g. an economic or climatic crisis) or cyclical events (e.g. seasonal food insecurity). The concept of stability can therefore refer to both the availability and access dimensions of food security.

Utilization: Utilization of food through adequate diet, clean water, sanitation and health care to reach a state of nutritional well-being where all physiological needs are met. This brings out the importance of non-food inputs in food security.

Source: FAO, 2006a.

Food security problems also arise when people lack knowledge about nutrition, food handling and preparation, lack access to clean water and sanitation or when their food supplies change and they have to deal with unfamiliar foodstuffs.

Every major conflict in history has destabilized local food supplies, often with wide ripple effects. So have crop and livestock pests and diseases, and natural disasters such as recurring drought in Ethiopia, annual floods in Bangladesh, earthquakes in Pakistan and Indonesia, and the 2010 fires that affected the Russian wheat crop. Fluctuating economic conditions drive vulnerable families below the poverty line and send them into food security crises, putting a strain on existing safety nets. For a middle class population with solid economic resources, a temporary rise in prices or a fluctuation in the food supply may be merely inconvenient – people must drive further to buy preferred foods, or divert a little more of their income into food purchases, or eat something different – but for vulnerable households, it creates a food security crisis.

Dimensions. The temporal dimensions normally refer to food insecurity, which can be **chronic**, resulting from a persistent shortage in supply or a systemic weakness that limits individuals' ability to access food, or **transitory**, arising because of a crisis. Both need to be addressed at the same time (Pingali *et al.*, 2005), because individuals and communities facing chronic food insecurity lack safety nets and are highly vulnerable to transitory problems, while an inappropriate response to a crisis may weaken the base for long-term food security by weakening local markets or creating dependencies. In 2005, the Committee on World Food Security (CFS, 2005) identified conflict as the most common cause of transitory food insecurity, followed by weather-related problems. In 2008 and 2009, the food security repercussions of the world economic crisis were a serious cause for concern (FAO, 2009a). As a result of transitory problems starting to blur into chronic food insecurity – mainly due to long-term systemic failures in the way that food is produced and distributed – the world is now facing the problem of protracted food crises (FAO, 2010a).

BOX 2
**DEALING WITH PROTRACTED FOOD CRISIS:
THE CASE OF ETHIOPIA**

In Ethiopia, where crop failure is an almost annual phenomenon, some 7 million people – more than 8 percent of the country's population – can support themselves from their own income for only six months a year. For the remaining six months, they rely on a recently introduced Productive Safety Net Programme that addresses the underlying structural problem of food insecurity by putting safety nets in place ahead of crises, such as by guaranteeing employment in public works for food or cash and direct subsistence payments. Supplying predictable disbursements of cash and food transfers at frequent intervals, as opposed to unpredictable disbursements at varying intervals, seems to have reduced the need to sell assets (especially livestock) to buy food, leaving people less prone to destitution from adverse weather events. However, even this could not provide ample protection from the soaring food prices and the drop in foreign investment and remittances that followed the 2007–08 economic crisis. (FAO, 2009a).

Long-term goals. The various pillars and dimensions of food security are encapsulated in two long-term goals that preoccupy the international community: **sustainable healthy diets** and **resilient food systems** (sometimes combined as sustainable and resilient food systems).

Sustainable healthy diets can be attainable if all of the conditions for food security are met in ways that do not unduly deplete natural resources or pollute the environment. "Sustainable" means that both present and future generations have sufficient food of adequate nutritional quality to promote their well-being (Pinstrup-Andersen, 2009; Harding, 2010). Under these conditions, food systems would have sufficient capacity to produce enough food of sufficient

variety consistently, transport it with minimum waste to where it is needed, provide it at prices that people can afford while also covering the costs of the externalities associated with food production, and to promote healthy choices in buying and preparing food. Currently, we face a growing population, finite fossil energy and water, and competition for land needed to produce human food, biofuel and livestock feed. For food systems to be sustainable, there is a need to address the structural and policy weaknesses that have contributed to creating the present situation.

Resilient food systems are those that withstand shocks from conflict, weather, economic crises, human or livestock diseases and crop pests. It is well recognized by relief agencies that their emergency aid efforts are most effective when they are injected into already resilient systems in ways that create minimum disruption. A more resilient global food system, therefore, would reduce the level and impact of transitory food insecurity. The food systems of developed countries are generally resilient, underpinned by strong economies and infrastructure, while those of most developing countries are not.

MEASURES

There is no one method to measure all of food security's dimensions, determine whether a food system is sustainable and resilient, and quantify the extent to which everyone in the world is consistently well nourished. Thus, it is necessary to rely on a range of measures that address the various aspects of food security.

The most direct, widely available and uniform measurement quantifies the consumption of calories – people who consume insufficient calories for their age and sex are considered undernourished. When the target was set in 1996 to halve hunger by 2015, there was already a promising trend in combating undernutrition. The number of undernourished people, standing at close to a billion in 1970, fell to 900 million in 1980, and to 845 million in 1990–92 (Table 1). Numbers stayed fairly static for the next ten years, rising

TABLE 1
NUMBER (MILLIONS) AND SHARE OF UNDERNOURISHED PEOPLE BY REGION 1990 TO 2007

COUNTRY GROUPS	1990–1992	1995–1997	2000–2002	2005–2007
World	843.4	787.5	833.0	847.5
Developed countries	16.7	19.4	17.0	12.3
	(2.0%)	*(2.5%)*	*(2.0%)*	*(1.5%)*
Developing World	826.6	768.1	816.0	835.2
	(98.0%)	*(97.5%)*	*(98.0%)*	*(98.5%)*
Asia and the Pacific	587.9	498.1	531.8	554.5
	(69.7%)	*(63.3%)*	*(63.8%)*	*(65.4%)*
Latin America and the Caribbean	54.3	53.3	50.7	47.1
	(6.4%)	*(6.8%)*	*(6.1%)*	*(5.6%)*
Near East and North Africa	19.6	29.5	31.8	32.4
	(2.3%)	*(3.7%)*	*(3.8%)*	*(3.8%)*
Sub-Saharan Africa	164.9	187.2	201.7	201.2
	(19.6%)	*(23.8%)*	*(24.2%)*	*(23.7%)*

Note: Percentages are share of total for the year.
Source: FAOSTAT.

slightly to 873 million in 2005. In percentage terms, the numbers were even more encouraging. In 1980, 28 percent of the world's population was undernourished. By 1990–92, the average had fallen to 16 percent for the world and 20 percent for developing countries, and in 2005–07 (the latest period for which comparable statistics are available), the figures stood at 13 percent for world population and 16 percent for developing countries (FAO, 2008a).

Since then, two global problems – increasing demand for biofuel and the world economic crisis – have created a serious block to halving hunger. Competition between food and fuel crops together with other factors resulted in rises in food prices in 2007 and the wider economic crisis that immediately followed reduced purchasing power. According to estimates, approximately 925 million people were undernourished in 2010, representing roughly 14 percent of the world's population of 6.8 billion. FAO databases show that undernourishment is unevenly distributed across regions, nations, households and individuals, with the main burden borne by the poorest countries and the poorest people.

Undernourishment is an important indicator of food insecurity, but it only tells part of the story. Food security is more than the consumption of sufficient calories; it is also about consuming food of adequate quality. People are malnourished if they eat insufficient calories or protein, food of poor quality, or if they are unable to utilize fully the food they eat (WHO, 2001). Diets can be poor if they lack minerals and vitamins, have insufficient fruits, vegetables or livestock products, or contain too much of elements that are harmful when taken in excess such as saturated fats and sugar (IFPRI, 2004). While 925 million people were undernourished in 2010, some 2 billion were estimated to be malnourished. Unlike undernourishment, which is associated with poverty, the problem of malnourishment is found across all income groups, although it takes different forms for the poor and the rich. The poorest lack an adequate supply of energy, protein and micronutrients, while for those who can afford sufficient calories, overconsumption and poorly balanced diets, together with their associated health problems, are an increasing problem (WHO, 2003).

BOX 3
COSTS OF MALNUTRITION

Preventing one child from being born with low birth weight in low-income countries was estimated to be worth US$580 in 2003 (Alderman and Behrman, 2003).

- In Nigeria, the annual economic loss due to malnutrition in children under five was estimated at US$489 million in 1994, or about 1.5 percent of GDP (FAO, 2004).
- In South Asia, the losses associated with iron deficiency have been estimated at US$5 million per year (Ross and Horton, 1998).
- In Bangladesh, the cost of iron deficiency in children has been estimated at nearly 2 percent of GDP (Ross and Horton, 1998).
- In India, elimination of child malnutrition would increase national income by US$28 billion. This is more than its combined expenditures for nutrition, health, and education.

- Diet-related chronic diseases were estimated to cost 2.1 percent of China's GDP in 1995 and 0.3 percent of Sri Lanka's (Popkin *et al.*, 2001).
- The cost of obesity has been estimated at 0.2 percent of GDP for Germany, 0.6 percent for Switzerland, 1.2 percent for the United States (WHO, 2007), 1 percent for Latin America and the Caribbean (PAHO, 2006), 1.1 percent for India, 1.2 percent for the USA and 2.1 percent for China (Yach *et al.*, 2006).
- The cost of diabetes has been estimated at 1.3 percent of GDP for the USA, 2.6 percent for Mexico and 3.8 percent for Brazil (Yach *et al.*, 2006).

Malnourishment is harder to measure than undernourishment, since it requires data on protein and micronutrients which are not routinely measured on a wide scale. Rough estimates can be made from the kilograms of different foods consumed and their average content of different nutrients. More commonly, malnourishment levels are deduced indirectly from proxy measures that show its resulting effects.

Malnutrition has a devastating effect on child survival, particularly in developing countries. It has been estimated that protein-energy malnutrition is a causative factor in 49 percent of the approximately 10.4 million annual deaths of children under five years of age (WHO, 2000). It is also manifested in underweight and stunting. In 2007, UNICEF estimated that approximately 146 million children were underweight (UNICEF, 2007), over 70 percent of them in developing countries, and that 31.2 percent of children in developing countries were stunted

(UNSC, 2010). This represented an improvement since 1980, when 49 percent of children under five in the developing world were stunted, and 38 percent were underweight (Opio, 2007).

At the other end of the scale, over-consumption can be deduced from statistics on obesity, defined as having a Body Mass Index (BMI), which measures body fat based on weight and height, of 30 or above. The most recent WHO global summary suggests that in 2008, at least 500 million adults were clinically obese (WHO, 2100), a figure which may rise to 700 million in 2015. Obesity is linked to diabetes and heart disease and possibly certain kinds of cancer.

Malnutrition not only affects an individual's health, it is expensive for society. It reduces human productivity and creates costs for the health system, as shown in Box 3. A conservative estimate in 1990 put the global economic loss from malnutrition at US$8.7 billion (Pinstrup-Andersen *et al.*, 1993).

The sustainability and resilience of food systems can be measured by a variety of qualitative and quantitative indicators, such as:

- trends in production and consumption levels per person as well as patterns of consumption among different income groups which give a general indication of resilience;
- long- and short-term trends in food prices and livestock disease prevalence which provide information about potential sources of food instability;
- information about water quality and other environmental indicators which provide underlying information about the resource base on which food production depends.

Key indicators can differ with individual national and local situations. For example, a country that relies on domestic production for the bulk of its food supply may be primarily concerned with measuring the ability of its own agricultural system to keep producing a stable supply or to store buffers against shocks, while a country that expects to import a proportion of its food every year will be equally concerned with the robustness of the international trade system and the political capital that gives access to food aid in times of crisis.

In rangeland areas of Africa, the terms of trade between livestock and food grains are an indicator of prolonged food emergencies because, as a crisis continues, more and more animals need to be sold to buy the same amount of grain. The EC-FAO Food Security Programme has developed a "resilience tool" to help policy-makers understand what makes families more resilient to crises. This tool combines several factors into an index, including income and access to food; assets such as land and livestock; social safety nets such as food assistance and social security; access to basic services such as water, health care and electricity; household adaptive capacity which is linked to education and diversity of income sources; and the stability of all these factors over time.

©FAO/L. Rlung

Livestock food in the diet

Animal source foods, a choice for many people in many societies, add taste, texture and variety to the diet. Some foods have specific social and cultural roles, such as turkey at Christmas, a duck taken as a gift on a social visit, eggs or milk given to lactating mothers, meat cooked for honoured visitors, tea with milk given to guests. Cultural norms also prohibit consumption of some foods, such as pork in Muslim and Jewish communities. Livestock contribute around 12.9 percent of global calories and 27.9 percent of protein directly through provision of meat, milk, eggs and offal, and also contribute to crop production through the provision of transport and manure.

NUTRITIONAL VALUE

In spite of recent growth in consumption, many people are still deficient in the nutrients that can be provided by animal source foods, which are complete, nutrient-dense and important for the high quality protein and bio-available micro-nutrients they contain, particularly for children

and pregnant and lactating women. Even quite small amounts of animal source foods are important for improving the nutritional status of low-income households. Meat, milk and eggs provide proteins with a wide range of amino acids that match human needs as well as bio-available micro-nutrients such as iron, zinc, vitamin A, vitamin B12 and calcium in which many malnourished people are deficient.

It is generally agreed that livestock source foods can be beneficial, but there are no universal guidelines that set an ideal level of consumption of livestock products for an individual. International dietary guidelines on levels of energy and protein consumption do not distinguish between plant and animal sources. They suggest that the intake of energy needed by an adult in a day varies from 1 680 to 1 990 kilocalories (kcals) in total, depending on the country. They also suggest that the safe level of protein consumption is about 58 g per adult per day. "Safe" in this case is defined as the average protein requirement of the individuals in the population, plus twice the standard deviation and it is an accepted practice to refer to this measure rather than a minimum (WHO, FAO, UNU, 2007).

TABLE 2

**AVERAGE DIETARY PROTEIN AND ENERGY CONSUMPTION AND
UNDERNOURISHMENT BY REGION**

COUNTRY GROUPS	PROTEIN CONSUMPTION *g/day 2003–05*	ENERGY CONSUMPTION *kcals/day 2005–07*	PERCENT OF POPULATION CONSUMING INSUFFICIENT **CALORIES** *2005–07*
World	76	2 780	13
Developed countries	102	3 420	<5
Developing World	70	2 630	16
United States of America	116	3 770	<5
Asia, the Pacific and Oceania	70	2 610	16
Latin America and the Caribbean	79	2 900	8
Near East and North Africa	83	3 130	7
Sub-Saharan Africa	53	2 240	28
Recommended "safe" consumption (adults)	58		
Minimum energy requirement		1 680–1 990	

Sources: FAOSTAT for all except "safe" consumption. Recommended "safe" consumption is estimated as the minimum average plus 2x standard deviation WHO, FAO, UNU (2007).

In most parts of the world, average consumption is above the minimum recommended level of energy and the safe level of protein, according to the most recent comparable consumption statistics. As shown in Table 2, only in sub-Saharan Africa is the average consumption of protein below the recommended safe levels. However, these averages hide a significant problem of malnutrition, with 16 percent of people in the developing world (28 percent in sub-Saharan Africa) estimated to be undernourished. Energy and protein consumption are quite closely linked, and insufficient calorie consumption tends to go in tandem with insufficient protein consumption.

These are average guidelines. Actual individual requirements depend on height, age, lifestyle and stage of life. Pregnant or lactating women, for example, need extra energy and protein. However, even the more detailed guidelines give only limited guidance about minimum requirements of livestock source food. National nutritional guides, such as those provided in the USA or the Netherlands, suggest including some livestock products in the diet but recommend that the largest proportion of food by weight should be in the form of fruit, vegetables and grains.

Excessive or inappropriate intake of livestock products creates risks and detrimental health effects. Increased consumption of red meats can increase the risk of colon cancer, and increased intake of saturated fats and cholesterol from meat, dairy products and eggs can increase the risk of chronic non-communicable diseases such as cardiovascular disease (UNSCN, 2005). National dietary guidelines typically warn against consumption of too much animal fat from meat and hard cheese and suggest a balance between livestock products and fish.

Since protein with a wide range of amino acids is a valuable dietary contribution from livestock, the range in livestock protein intake levels according to geographic area is worth examining. Table 3 shows that the consumption per person of livestock protein increased in all areas of the world between 1995 and 2005. However, it also shows that average consumption in Africa remained at less than a quarter of that in the Americas, Europe and Oceania, and Africa's livestock protein consumption was a modest 17 percent of the recommended safe level for all proteins. By contrast, the consumption of livestock protein in the Americas, Europe and Oceania in

TABLE 3

AVERAGE DAILY CONSUMPTION PER PERSON OF LIVESTOCK PROTEIN COMPARED TO SAFE LEVEL 1995 AND 2005

| | | G/DAY | | | | % OF RECOMMENDED "SAFE"[1] CONSUMPTION FROM LIVESTOCK |
| | | MEAT | DAIRY (NOT BUTTER) | EGGS | TOTAL | |
AREA	YEAR					
Africa	1995	5.3	3.1	0.6	9	
	2005	5.9	3.4	0.6	9.9	17
Americas	1995	26.1	14.3	2.7	43.1	
	2005	28.1	14.1	3.1	45.3	78
Asia	1995	7.5	3.8	2.2	13.5	
	2005	9.2	4.7	2.7	16.6	29
Europe	1995	24.1	17.9	3.6	45.6	
	2005	24.7	19.2	3.8	47.7	82
Oceania	1995	24.9	18	1.9	44.8	
	2005	39.3	15.8	1.7	56.8	98
Least developed countries	1995	3.3	2.2	0.2	5.7	
	2005	4.1	2.7	0.3	7.1	12

Source: FAOSTAT for consumption figures.
[1] Recommended "safe" consumption is 58 g per person per day, estimated as the minimum average plus 2x standard deviation (WHO, FAO, UNU, 2007).

2005 was between 78 and 98 percent of the total protein requirement, suggesting that livestock products were being over-consumed. The high level of meat and saturated fat consumption in high-income countries has been associated with high rates of cardiovascular disease, diabetes and some cancers (Walker, 2005).

Even in small amounts, food of animal origin can play an important role in improving the nutritional status of low income households by addressing micro- and macronutrient deficiencies, particularly of children and pregnant and lactating women. It is possible to live healthily without eating animal products, but they do provide nutritional benefits, particularly through micronutrients. Small amounts of meat, for example, provide easily absorbable haem iron and help in the absorption of iron from plant foods (Bender, 1992), which helps prevent anaemia arising from iron deficiency. Meat and milk are good sources of vitamin B12, riboflavin and vitamin A. Meat also provides zinc, and milk provides calcium. Adding a small amount of animal source food to the diets of malnourished children can increase their energy and cognitive ability (Neuman *et al.*, 2010). However, it is important that babies receive human milk up to age 6 months, rather than a livestock source substitute (Neuman, 1999). Iron deficiency, for example, is estimated to affect 1.6 billion people worldwide (deBenoist *et al.*, 2008) and to impair the mental development of 40–60 percent of children in developing countries (UNICEF, 2007). A multi-agency report in 2009 stated that iron deficiency anaemia during pregnancy is associated with one-fifth of total maternal deaths each year. (Micronutrient Initiative, 2009). Meat is not the only source of dietary iron, but it is a good source. It seems clear that the poor would benefit from a higher intake of food, with a diet that includes livestock source food. Therefore, the next section examines the sources of animal products in the diets of poor households.

LIVESTOCK PRODUCTS IN THE DIETS OF THE POOR

Poorer households spend less than richer ones on food, particularly on livestock food items. This topic is dealt with in considerable detail in a later chapter, which looks at food access, but it is worth mentioning a few statistics here. National consumption figures show that consumption of livestock source foods rises as average income rises (Delgado, 2003), which is illustrated in Figure 6. Studies within individual countries also show differences between rich and poor households. For example, a comparative study of Uganda, India and Peru (Maltsologu, 2007) found that poor households consumed less in both volume and total value of livestock products than rich ones, with the poorest households allocating less than 10 percent of their food budget (purchases and home consumption) to livestock products. Within the budget given to livestock source foods, the highest percentage was allocated to meat. In Uganda, milk was also important, while eggs were more prominent in Viet Nam.

While there are differences in food preferences and access between countries and even within households (described later when reviewing food access), both poultry products and dairy products tend to be prominent within the diets of poor households.

Poultry meat and eggs. Globally, the supply of and demand for poultry products has shown a very rapid upward trajectory, with poultry now providing 28 percent of all meat (see the next chapter for trends in livestock production). Poultry meat and eggs are acceptable foods in many cultures, and poultry can be raised at home even by families with very little land or capital, making them easily accessible to the poor. In some countries, poultry meat is cheaper, such as in Egypt where at times, it is only one-third the price of other meats (Hancock, 2006). Poultry products make up 0.6 percent of the average of 2 077 kcals per person per day in Africa and 2.9 percent of 2 300 kcals per person per day in Asia (Hancock, 2006). They form a somewhat

greater share of average protein consumption, up to 5 percent in the poorest households. Anecdotal and recorded evidence points to poultry products contributing over 20 percent of meat consumption in sub-Saharan Africa, around 50 percent in Egypt and the more food insecure countries of Latin America, and a high percent in the poorer Middle Eastern countries. This makes these populations particularly vulnerable when local production of poultry is disrupted because of disease or other problems.

Quickly cooked and digested, poultry meat and especially eggs also have good micronutrient properties important for children and pregnant women. Female-earned income from poultry keeping is an important factor for improved child health, and smallholder poultry development projects for poor households in Bangladesh and South Africa indicate that both the direct consumption of poultry and income from poultry contribute to reduced malnutrition (Dolberg, 2003).

Dairy. Milk from cattle and goats, a good source of amino acids and Vitamin A, is widely consumed in all parts of the world except East Asia. In South Asia, Africa and the Middle East, it is particularly important in the diet and, in fact, can contribute more than 50 percent of pastoralist families' energy intake. School milk programmes have been used to boost consumption

©FAO/Ivo Balderi

by children while supporting the local dairy industries. A 2004 survey of 35 countries found that schemes to promote milk consumption in schools had increased the proportion of school milk in the domestic market. In Thailand, where milk is not a large part of the national diet, school milk accounted for 25 percent of national milk consumption, while in other countries that responded to the survey, the contribution was between 1 and 9 percent (Griffin, 2004).

Smallholder dairy production also has been important to rural economies, although not to the poorest of the poor since the maintenance of a cow or even a dairy goat is normally beyond their capacity. Women often have control of dairy animals and of the income they provide, which has had positive consequences for household nutrition, a topic that is explored further in a later chapter reviewing food access.

©FAO/Florita Botts

Livestock and food supply

Livestock products supply around 12.9 percent of calories consumed worldwide (FAO, 2009b) and 20.3 percent in developed countries. Even more important, perhaps, is their contribution to protein consumption, estimated at 27.9 percent worldwide and 47.8 percent in developed countries.

SUPPLY OF ANIMAL SOURCE FOODS
The availability of livestock products worldwide and within nations is determined by the volume of production and the scale and reach of international trade. During the past 40 years (1967–2007), global production of meat, milk and eggs has grown steadily. Particularly striking have been the increases in production of poultry meat by a factor of 7.0, eggs by a factor of 3.5, and pig meat by a factor of 3.0 (Table 4). Production per person has also grown, albeit at a slower rate. For the decade from 1995 to 2005, the annual global growth rate in consumption and production of meat and milk averaged between 3.5

and 4 percent, double the growth rate for major staple crops during the same period (Ahuja *et al.*, 2009). Trade in livestock products also has grown enormously during these 40 years (Table 5), by a factor of 30.0 for poultry meat, more than 7.0 for pig meat and 5.0 for milk.

While the global supply of livestock products has more than kept up with the human population expansion, the situation has not been the same in all regions. Production levels have expanded rapidly in East and Southeast Asia, and in Latin America and the Caribbean, but growth in sub-Saharan Africa has been very slow. Fast growth in human populations in some developing countries coupled with low productivity per animal have made it hard for livestock production in those areas to keep up. There is also considerable variation within the developing world, with sub-Saharan Africa and South Asia producing at much lower levels per person than Latin America and the Caribbean.

Pigs and poultry, especially those kept in intensive, peri-urban production systems, are mostly responsible for per person growth of livestock source foods. Three of the largest emerging economies – China, Brazil and India –

TABLE 4

CHANGES IN GLOBAL LIVESTOCK PRODUCTION TOTAL AND PER PERSON 1967 TO 2007

ITEM	PRODUCTION *(million tonnes)*			PRODUCTION PER PERSON *(kg)*		
	1967	2007	2007/1967	1967	2007	2007/1967
Pig meat	33.86	99.53	294%	9.79	14.92	152%
Beef and buffalo meat	36.50	65.61	180%	10.55	9.84	93%
Eggs, primary	18.16	64.03	353%	5.25	9.60	183%
Milk, total	381.81	680.66	178%	110.34	102.04	92%
Poultry meat	12.39	88.02	711%	3.58	13.20	369%
Sheep and goat meat	6.49	13.11	202%	1.88	1.97	105%

Source: FAOSTAT.

have fast-growing poultry industries (Figure 1). China is by far the largest player, with approximately 70 million tonnes of egg production annually compared to 3 million tonnes in India and 2 million in Brazil, and 15 million tonnes of meat compared to 9 million tonnes in Brazil and 0.6 million in India. However, poultry make an important contribution to the food supply in all three economies. In India, poultry is the fastest growing livestock subsector. Poultry products accounted for approximately 50 percent of per person livestock protein consumption in 2003, compared to about 22 percent in 1985 (Pica-Ciamarra and Otte, 2009, based on Government of India, 2006). China and Brazil are also rapidly expanding their production of pig meat (Figure 1). In China in particular, this is an important part of the diet.

Dairy production has expanded to meet demand in some growing economies of Asia, such as in Thailand where domestic dairy production rose sharply from 7 percent of national consumption in 1980–82 to 44 percent in 2000–02 (Knips, 2006). Viet Nam, which only has a short national tradition in the production and consumption of dairy products, saw a tripling in milk production between 1996 and 2002 (Garcia *et al.*, 2006). Although Pakistan still faces milk shortages, because of limited feed and grazing areas coupled with a rising population, farmers

TABLE 5

CHANGES IN GLOBAL TRADE OF LIVESTOCK PRODUCTS 1967 TO 2007

ITEM	EXPORT *(million tonnes)*		
	1967	2007	2007/1967
Pig meat	1.48	11.13	750%
Beef and buffalo meat	2.41	9.46	392%
Eggs, primary	0.33	1.44	442%
Milk, total	18.84	93.19	495%
Poultry meat	0.39	12.66	3 206%
Sheep and goat meat	0.58	1.04	180%

Source: FAOSTAT.

have responded to increased demands for milk by increasing milk yields (Garcia *et al.*, 2003). In India, where milk has always been important, the latest statistics from the National Dairy Development Board (NDDB) show that availability per person has grown from 178 g per day in 1991–92 to 258 g per day in 2008–09 (NDDB, 2010).

Many poor countries, however, have failed to increase national production or consumption of livestock and livestock products. In Bangladesh, for example, high milk production costs and low yields have resulted in low per person

1 PRODUCTION FROM POULTRY AND PIGS IN INDIA, BRAZIL AND CHINA 1967 TO 2007

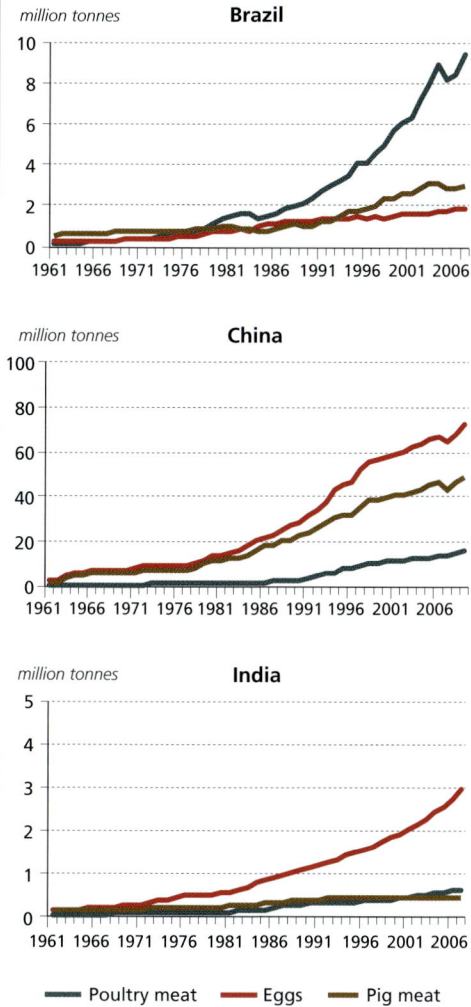

million tonnes — **Brazil**

million tonnes — **China**

million tonnes — **India**

— Poultry meat — Eggs — Pig meat

Source: FAOSTAT.

consumption per person of livestock products (Halderman, 2005).

Export trade, which in 1967 was relatively small and dominated by Europe, has not only expanded greatly, it has diversified, with the Americas becoming the dominant exporter of poultry meat, Asia taking a growing share of egg and poultry meat trade, and Oceania showing strong growth in milk and ruminant meat exports (Figure 2).

There is a large gap in self sufficiency in livestock products between the developed and the developing regions. Oceania is a major net exporter of ruminant meat and milk, including exports of live sheep, many to the Middle East and North Africa. The Americas are increasingly net exporters of pig and poultry meat, Europe is self sufficient in some products and a minor net importer of others, and Africa is a net importer of almost all livestock products (Figure 3).

Within regions, some countries stand out as major producers and net exporters while others are net importers and rely on trade to make livestock products available in their domestic markets. For example, Asia as a whole is barely self sufficient in poultry meat, but Thailand has been among the top ten exporters, and China is a major producer with a growing export market. Within the Americas, the USA and Brazil stand out as exporters of livestock products while some of the smaller countries are net importers. The biggest milk powder importers are oil exporters such as Mexico, Algeria, Venezuela and Malaysia, and the fast-growing economies of India, the Philippines and Thailand (Knips, 2005). In China, domestic milk production has risen but still has not been able to keep up with rising demand as domestic milk consumption has increased even faster. As a result, milk powder imports have risen rapidly to meet demand. North Africa, which has experienced rapid income growth in the past few years, has become a large importer of milk powder to meet increased demand for dairy products.

Global availability of livestock products has grown, but how close does it come to what is

milk production of 13 kg per year. Even with imports, the country is struggling to meet a domestic milk demand that has increased as a result of rising incomes and population growth (Garcia *et al.*, 2004a). Ethiopia, which has one of the largest livestock populations in Africa, has seen a decline over the past 30 years in the number of livestock and the volume of livestock production per person, and a corresponding decline in

2 EXPORT OF LIVESTOCK PRODUCTS BY REGION 1967 AND 2007

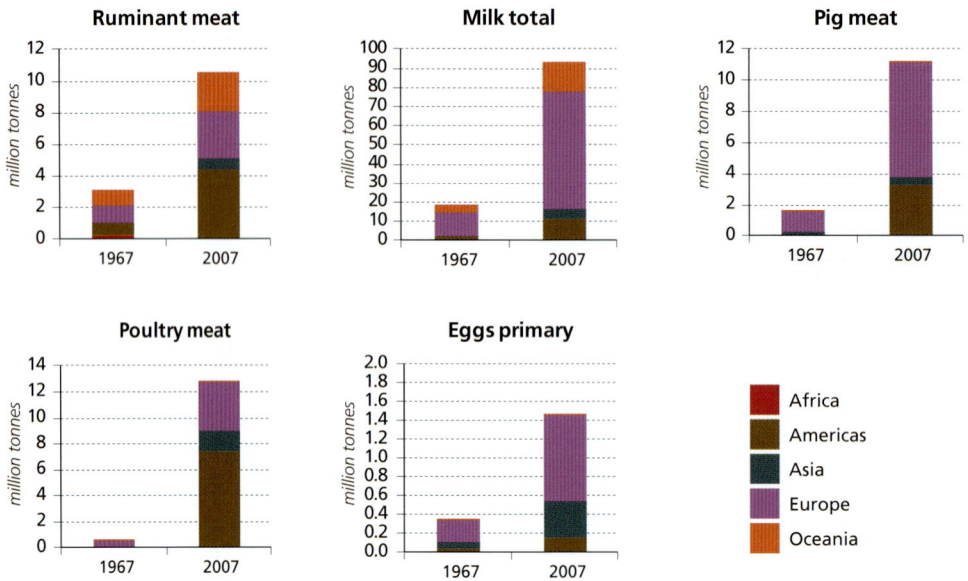

Ruminant meat

Milk total

Pig meat

Poultry meat

Eggs primary

Africa
Americas
Asia
Europe
Oceania

Source: FAOSTAT.

needed for food security? The literature tends to compare developed to developing country statistics rather than comparing the consumption in the developing world to acceptable nutrition standards. Perhaps this is because the question has no simple answers. The recommended standards for consumption of calories, proteins and certain critical micronutrients do not generally distinguish between the sources of food, other than to say that a balanced diet should contain a mixture of nutrients from plant and animal sources with a higher proportion coming from plant sources.

Expert opinion suggests that sufficient food of all kinds is currently being produced for everyone, but that the problem lies with access. At the same time, cognizant of the fact that food security requires a sufficient supply of both crop and livestock products, it is important to examine the interplay between crop and livestock production. They interact in both positive and negative ways. In mixed farming systems, the two add value to each other – livestock provide traction and manure for crop production and crops, in return, provide forage and residues as livestock feed. A tug-of-war develops when livestock consume grains and other seeds that could otherwise be fed to humans and, in doing so, compete for the food needed for direct human consumption.

LIVESTOCK CONTRIBUTING TO CROP PRODUCTION

In addition to contributing directly to food supply through provision of their own products, livestock contribute indirectly by supporting crop production with inputs of manure and traction. In both cases, their contribution is greatest in developing countries. In the developed world, the use of traction has fallen to almost nothing, and the manure produced by livestock raised for food is more than can be used conveniently on local cropland.

3 NET TRADE PER PERSON IN LIVESTOCK PRODUCTS BY YEAR AND REGION

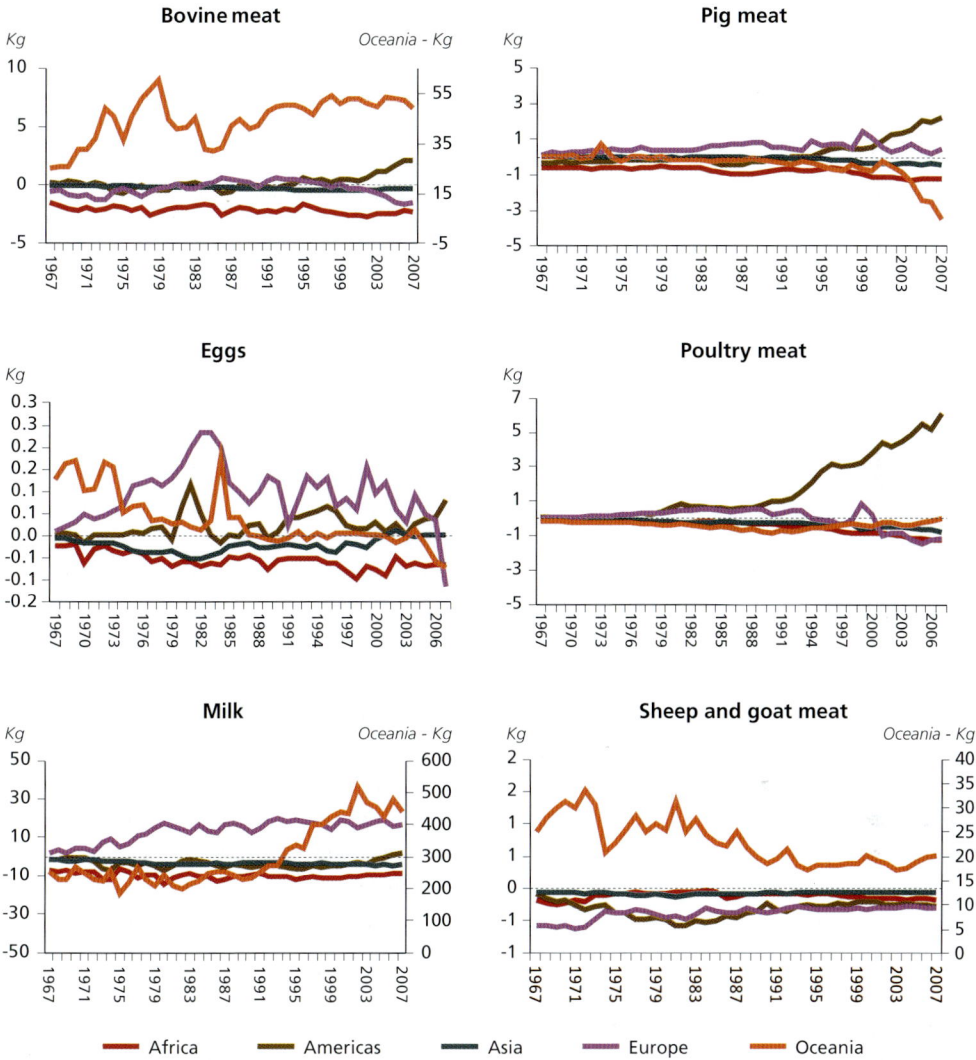

Source: FAOSTAT.

DRAFT POWER

Draft power from working animals has reduced human drudgery, allowed cropping areas to be expanded beyond what can be cultivated by hand, and made it possible to till land without waiting for it to be softened by rain which gives farmers more flexibility in when they plant crops. In spite of this, a recent review (Starkey, 2010) indicates that the number of working animals in the world has probably fallen from 300–400 million in the 1980s to 200–250 million today.

Numbers in Africa have increased (Box 4), but there have been significant decreases in other parts of the world. In Western Europe and North America, the use of animal power has almost disappeared since WW II other than for

BOX 4
EXPANSION OF ANIMAL TRACTION IN AFRICA

West Africa – animal traction continued expanding during the twentieth century, due to its promotion by commodity companies and extension services. There have been high levels of adoption in the 400–800 mm rainfall zone, and use of work oxen in francophone West Africa increased six fold – from 350 000 to 2 million – in the past 50 years. Oxen are the main agricultural work animals, but horses and donkeys are also used in the drier areas. Donkeys have increased in numbers, from 4.5 to 6.3 million in the past decade, and in geographical area, with the "donkey line" moving southward. In the humid zone, there are few cattle and no equids, but projects are considering the introduction of work oxen. An increasing number of farmers use trypanotolerant Ndama cattle for work in Guinea.

East Africa – animal traction is gradually increasing, notably in Tanzania, with 1 million work animals, and in Uganda, while in Madagascar, where 300 000 ox carts are in use for transport, bovine traction was badly affected by the 2006 drought. Animal use is slowly diversifying from the traditional ploughing and pulling of carts to increased use for weeding and conservation tillage and increased use of donkeys for transport and light tillage.

Ethiopian highlands and some neighbouring areas – 7 million oxen provide the main source of power for soil tillage, while 5 million donkeys are used for pack transport. Donkey carts are few but increasing. Horses and mules are used widely for riding, although urban horse carts are being replaced by motorized three-wheelers. In Ethiopia, ox-drawn ploughing is so important that poor households that do not own oxen will practice sharecropping with those that do and give as much as 50 percent of their harvest in exchange for the use of oxen (Ashley and Sandford, 2008).

Southern Africa – animal traction has been in use since the seventeenth century, making it traditional in many smallholder systems. In recent decades, it has been promoted and is spreading in several countries, including Malawi, Namibia and Zambia.

South Africa and neighbouring countries – the use of tractors on large farms and subsidized tractor hire schemes have diminished people's perception of the value of animal traction. However, no viable system for using tractors for rainfed crops on fragmented small-scale farms has been found. Oxen are the preferred animal for ploughing, but droughts, overgrazing and theft have made donkeys more attractive.

North Africa – traditional use of work animals in agriculture remains important in Egypt and Morocco.

Source: Starkey (2010), except where indicated.

---

specialized uses and in traditional communities, such as the Amish in North America. In Eastern Europe, it is steadily decreasing as tractors become more affordable and available and farm sizes shrink.

In much of South and Southeast Asia, draught animals are being replaced by mechanization. In Central and South America, oxen and horses remain common on smallholder farms in spite of increasing adoption of tractors, and animal-drawn carts are quite widely used for rural and urban transport. The traditional use of pack llamas has declined greatly but donkeys remain important in the Andes and in Mexico. Animal traction also remains important for agriculture and transport in Haiti and the Dominican Republic, although motorcycles, three-wheelers and power tillers may eventually reduce de-

4 ARABLE LAND AREA BY YEAR AND REGION

Thousand ha

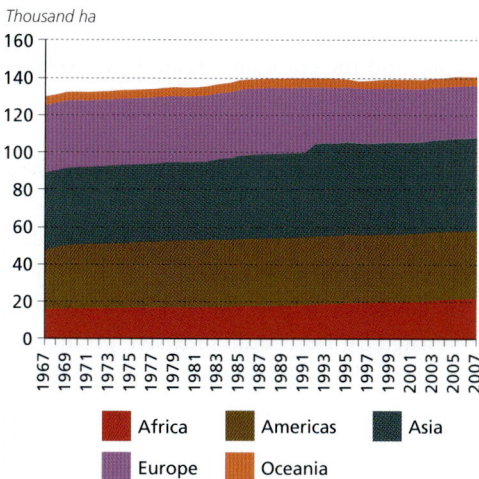

Source: FAOSTAT.

such as sub-Saharan Africa, as well as the persistence of animal power in both poor and rapidly industrializing countries, and the stability of some donkey populations. However, people replace animals when motor power is available, affordable, profitable and socially acceptable. Young people influenced by media images may consider animal power too old-fashioned to be socially acceptable. Also, with the exception of a few African countries, government support to research, education, training and promotion in the use of animal traction has declined.

The implications of the trends are complex. On a national scale, animal traction may be less energy efficient than mechanical tillers (Sharma, 2010), and there may be no incentive for many governments to promote it. Work animals also have their drawbacks. They need to be fed and cared for daily, they are vulnerable to disease and theft, they need feed either grown or purchased, they require specialist expertise, and they may be seen as old fashioned by young people. Tractors also increase labour productivity, giving some family members the option to migrate to cities.

Against these drawbacks must be set the very important role that work animals play in the lives and livelihoods of many families, particularly those that are poor or live in remote or hilly areas. Where animal traction is growing, increased farm power, crop-livestock integration and transport capacity should lead to greater, more stable production, marketed produce and incomes. Replacing animals with tractors can increase soil compaction and reduce manure availability for fertilizer or fuel, while tractors seldom increase yields per hectare (Starkey, 2010). As climate change is associated with higher frequencies of extreme weather, transport animals such as donkeys may prove increasingly important for access following natural disasters.

Animal traction is resilient even without a supporting policy environment, and the existing trends will generally continue, with areas of decline, stability and slow growth. However, as fewer people learn about work animals, it will

mand. Throughout the world, even in countries where the number of work animals is falling, pockets of use remain in remote and poor communities, where livestock make an important contribution to livelihoods.

The land devoted to crops increased globally by less than ten percent between 1967 and 2007 (Figure 4), although cropland locations shifted as cities expanded and forests contracted or expanded. The proportion of the world's cropland located in Asia and the Americas grew slightly while that in Europe decreased (Figure 4). This means that trends in the use of draft animals do not depend on a growing cropland base but rather on factors such as comparative costs and convenience of power tillers and tractors, farm size and remoteness, social custom and policies that support or depress the use of work animals (Starkey, 2010).

People will replace human-powered tillage and transport with animals when they are available, adapted to the environment, affordable, profitable and socially acceptable, and where no viable mechanization alternatives are available. This explains the animal traction growth in areas be more difficult to formulate appropriate policies relating to their use in agriculture and transport. A reasonable level of public investment in animal traction will need to be maintained for farmers in zones where such technology can directly reduce poverty and drudgery. Building a critical mass of knowledgeable users and support services, however, generally requires project support.

MANURE

The potential contribution of animal manure to crop production is well understood although there is no convenient global database to summarize its current contribution. It is easier to determine the extent of artificial fertilizer use, which is expected to double in developing countries by 2020 (Bumb and Baanante, 1996). In developed countries, it has been suggested that only about 15 percent of the nitrogen applied to crops comes from livestock manure. In developing countries, the relative contribution of livestock manure can be high but is not well documented.

The relationship between manure and food production is interesting and complex. It is a valuable input, but also a comparatively inconvenient one. Manure is known to be better than artificial fertilizer for soil structure and long-term fertility. Its greatest value can be seen in developing countries, where small-scale farmers report that they do not have enough manure to apply to their crops (Jackson and Mtengeti, 2005) and exchange of grain and manure occurs between settled farmers and pastoralists (Hoffman *et al.*, 2004). The distance that manure is sometimes transported attests to its perceived value. For example, chicken manure is reportedly transported 100 km or more in Viet Nam. Calculations made for Bolivia point to the considerable potential benefits of using more of the national manure production as an input to small-scale cropping (Walker, 2007). It also has multiple uses for household fuel, construction and biogas production as well as fertilizer, although these are not being fully exploited. One

estimate suggests that only 1 percent of global manure production is recycled as biogas (Thøy *et al.*, 2009). At the same time, it is less convenient to handle than artificial fertilizer, has variable quality, and the reduction in animal traction in many countries has also reduced the availability of this resource. Research into rice production in Asia, where work animals have been replaced by tractors and power tillers, is increasingly centered around more efficient ways to formulate and deliver artificial fertilizer.

In countries where the livestock sector is dominated by large-scale intensive production, manure can be as much a problem as a benefit. The challenge of recycling waste in ways that do not add to water pollution is substantial (Steinfeld *et al.*, 2006). For example, the EU and Canada (Hofmann, 2006) have strict rules and detailed guidelines about storage, processing and application of animal waste to avoid pollution of runoff water and the build-up of heavy metals in the soil. Denmark has successfully reduced leaching intensification, and concentration of its livestock sector has resulted in more manure being generated in smaller areas. For example, Figure 5 shows the large increase in manure production expected from increasing commercialization of the Vietnamese poultry sector in a country where poultry manure is already transported over quite long distances.

The extent to which livestock manure is applied to crops is a question of economics, logistics and regulation. There is evidence that using manure on small- to medium-sized mixed farms has economic viability (Bamire and Amujoyegbe, 2004). However, storage needs, transport requirements and the relative locations of livestock and crops all affect the cost and convenience of applying manure, as do government regulations on nutrient management (Kaplan *et al.*, 2004). Much current research is focussed on ways to tighten nutrient cycles, so that more nitrogen (N) and phosphorus (P) cycle through plants and animals and less is lost. In other words, the goal is to use more of these nutrients directly in agriculture (Steinfeld *et al.*, 2010).

5 PROJECTIONS FOR CHICKEN MANURE
PRODUCTION IN VIET NAM WITH
CHANGES IN SECTOR STRUCTURE

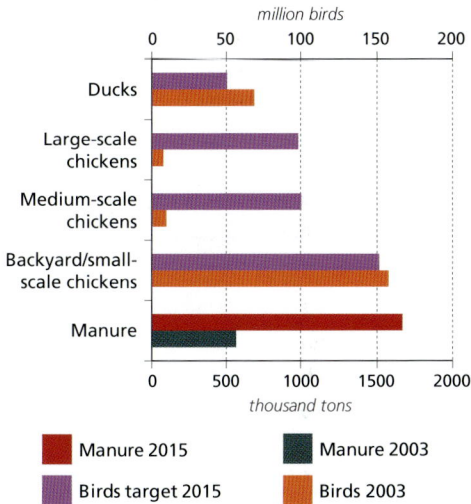

Source: Hinrichs, 2006.

LIVESTOCK AND THE FOOD BALANCE

Livestock make their most important contribution to total food availability when they are produced in places where crops cannot be grown easily, such as marginal areas, or when they scavenge on public land, use feed sources that cannot directly be eaten by humans, or supply manure and traction for crop production. In these situations, they add to the balance of energy and protein available for human consumption. When livestock are raised in intensive systems, they convert carbohydrates and protein that might otherwise be eaten directly by humans and use them to produce a smaller quantity of energy and protein. In these situations, livestock can be said to reduce the food balance.

In a world that is increasingly concerned with sustainable food production, ideally the contribution of livestock to the food balance should be at least neutral, making the conversion of natural resources to human food as efficient as possible while also ensuring that people still have the possibility of eating a diverse diet that includes livestock products. However, on a global scale, this is not the case and may not even be possible. It is estimated that 77 million tonnes of plant protein are consumed annually to produce 58 million tonnes of livestock protein (Steinfeld *et al.*, 2006).

The production system and the species of livestock both affect the food balance. Monogastrics such as pigs and poultry naturally eat a diet that is closer to a human one than that of ruminants. Extensive systems require animals to find a large proportion of their feed from sources not edible to humans, such as grasses and insects, grains left over from harvests and kitchen waste, while animals in intensive systems are fed concentrate feed that includes cereals, soya and fishmeal as well as roughage. Intensive poultry and pigs are the biggest consumers of grain and protein edible by humans, although both have been bred to be efficient feed converters. Intensive beef systems in feed lots convert concentrates less efficiently but can be fed partly on brewers' waste. Intensive dairy cows are fed concentrates that enable them to produce much greater volumes of milk than they could manage from a roughage-only diet.

The systems that compete least for human food – those that primarily depend on grazing – produce only about 12 percent of the world's milk and 9 percent of its meat. Mixed systems in which animals eat grass and crop residues as well as concentrates produce 88 percent of the world's milk and 6 percent of its meat. The most intensive industrial livestock systems are termed "landless" because the animals themselves occupy little land – they are kept in controlled environments and can be housed almost anywhere. These systems (Table 6) produce 45 percent of the world's meat, much of it from poultry and pigs, and 61 percent of the world's eggs (FAO, 2009b).

Since livestock have an important role in protein production, it serves as a valuable exercise to consider the effect of livestock production systems on the available balance of human-edible protein. This report makes an initial attempt to compare national figures for livestock output

TABLE 6
GLOBAL LIVESTOCK PRODUCTION AVERAGE BY PRODUCTION SYSTEM 2001 TO 2003

| | LIVESTOCK PRODUCTION SYSTEM | | | | |
	GRAZING	RAINFED MIXED	IRRIGATED MIXED	LANDLESS/ INDUSTRIAL	TOTAL
			(Million head)		
POPULATION					
Cattle and buffaloes	406	641	450	29	1 526
Sheep and goats	590	632	546	9	1,777
			(Million tonnes)		
PRODUCTION					
Beef	14.6	29.3	12.9	3.9	60.7
Mutton	3.8	4.0	4.0	0.1	11.9
Pork	0.8	12.5	29.1	52.8	95.2
Poultry meat	1.2	8.0	11.7	52.8	73.7
Milk	71.5	319.2	203.7	-	594.4
Eggs	0.5	5.6	17.1	35.7	58.9

Source: Steinfeld *et al.*, 2006

and feed input for a selection of countries. Using FAOSTAT production and trade statistics and feed and primary crop data, the estimated volume of edible livestock produced in each country has been adjusted for protein content of each commodity and then compared with the estimated volume of human edible protein that has been used for feed (domestically produced and imported). The input and output figures have then been compared as net figures and ratios, shown in Table 7. The numbers need to be treated with some caution, as feed data are somewhat limited and likely to underestimate the use of feed that is produced on small farms. However, the trend fits with what common sense might suggest: the countries with the most concentrated and intensive systems have an output/input ratio of below or near one (1), meaning that the livestock sector consumes more human-edible protein than it provides, while those countries with a predominance of extensive ruminants have considerably higher ratios, meaning that they add to the overall supply of protein.

Reducing the amount of human-edible food needed to produce each kilogram of livestock source food processed through livestock would be a valuable contribution to food security. There are two ways that this might be done: i) produce a larger percentage of the world's livestock protein within grazing and low intensity mixed systems, leaving more plant protein to be eaten by humans, or ii) recycle more waste products, including agro-industrial by-products, through animals. Both of these possibilities will be examined under "Producing enough food". There is no single approach to producing sufficient livestock source foods in a sustainable way. Rather than making blanket recommendations about livestock production, there is a need to balance the food security needs of the different human societies, a discussion also explored later in the report.

STABILITY OF FOOD SUPPLIES
Food security can be compromised when crops and livestock are destroyed or market chains disrupted, cutting off supplies, or when economic crises or loss of livelihoods abruptly reduce

TABLE 7

**HUMAN-EDIBLE PROTEIN BALANCE IN THE LIVESTOCK PRODUCTION OF
SELECTED COUNTRIES**

	EDIBLE PROTEIN OUTPUT/INPUT		EDIBLE PROTEIN OUTPUT–INPUT TONNES	
	AV. 1995–1997	AV. 2005–2007	AV. 1995–1997	AV. 2005–2007
Saudi Arabia	0.15	0.19	-533 731	-659 588
USA	0.48	0.53	-7 846 859	-7 650 830
Germany	0.66	0.62	-921 449	-1 183 290
China	0.75	0.95	-2 822 998	-665 276
Netherlands	1.66	1.02	322 804	18 070
Brazil	0.79	1.17	-622 177	550 402
Nepal	2.25	1.88	37 370	40 803
India	3.60	4.30	2 249 741	3 379 440
Sudan	18.22	8.75	235 868	340 895
New Zealand	8.04	10.06	460 366	638 015
Mongolia	14.72	14.60	42 987	35 858
Ethiopia	16.02	16.95	99 909	141 395
Kenya	18.08	21.16	124 513	202 803

Original data: FAOSTAT, November 2010. Calculations by FAO Animal Production and Health Division.
Edible protein output estimated from indigenous meat, milk and eggs. "Indigenous" meat production = production from slaughtered animals plus the meat equivalent of live animal exports minus the meat equivalent of live animal imports.
Edible protein input estimated from available feed (domestically produced and imported) and primary crops that are edible by humans (excluding canary seed and vetches).

access to food. Wars and conflicts, economic crises, fires, floods, droughts, earthquakes, tsunamis and major epidemic diseases have all destabilized food security, sometimes affecting both supply and demand (Box 5). Long global food chains and the dominance of some exporting countries mean that local problems can have regional or global effects (Stage *et al.*, 2010). Resilient food systems have inbuilt factors that help stabilize them or help them recover from instability. Livestock contribute in a number of ways to the food stability of their owners and the nations where they are produced. However, they are vulnerable to disease and natural disasters and, if these effects are not addressed, the beneficial effect of livestock on the stability of food supplies will be reduced.

LIVESTOCK AS A BUFFER
Livestock represent part of a family's risk management strategy. Building an economic and so-

cial buffer against shocks is an important part of ensuring food stability. It is well known that families below or near the poverty line are particularly vulnerable to shocks since they already devote a large proportion of their income and resources to securing food and have very little margin to cope with extra stress. Livestock are an asset that can help to build these buffers. They grow and reproduce, providing an expanding asset base for their owners. Herd accumulation is a common practice even among agropastoralists, for whom livestock represent a minor income source during normal times (Ashley and Sandford, 2008). Several years of crop failure in Pakistan motivated farmers to increase their livestock numbers, in order to manage risk through diversification (Garica *et al.*, 2003). Very poor landless urban dwellers also may keep a few small livestock as a buffer against risk. A 2003 study in Uganda found that livestock ownership in Kampala increased dur-

BOX 5
NATURAL AND ECONOMIC SHOCKS TO FOOD SYSTEMS

Natural: El Niño events

El Niño events are weather events that usually take place every four to seven years and last for one or two years. Recently, they have been occurring more frequently, causing flooding in some parts of the world and drought conditions in others, resulting in loss of crops, livestock, infrastructure and property as well as displacement of people. El Niño events are a particular concern because their effects are unpredictable and it is hard to take preventive action.

The 1987–88 El Niño caused massive flooding in 41 countries along the coast of Latin America and in parts of the Horn of Africa, droughts or dry spells in Southeast Asia and major forest fires in Indonesia and Brazil, with a total cost of between US$32 billion and US$96 billion. In Indonesia, drought caused a shortfall of over 3.5 million tonnes in the cereal harvest, and food prices rose sharply. In Somalia, harvests stored underground were destroyed by flooding There were considerable losses of livestock in Kenya, Somalia and Ethiopia due to unseasonable and heavy rainfall and floods, as well as an outbreak of the zoonotic disease Rift Valley fever (RVF) in Kenya and Somalia. In Southern Africa, El Niño tends to cause prolonged dry spells during the period between January and March when rainfall is most required by crops, resulting in reduced yields or, in some cases, complete crop failure as well as reduced output from pastures. The prices of staple foods rise, livestock conditions deteriorate and livestock prices decline when households make emergency sales of animals to meet household expenses.

Economic: global economic crisis

The economic crisis of 2007–2008 produced unusually rapid food price increases when the rising cost of energy had knock-on impacts on food production costs, creating a livelihoods shock for poor families. In 2007, increases in the number of undernourished people occurred in Asia and the Pacific and in sub-Saharan Africa, the two regions that together accounted for nearly 90 percent of the undernourished people in the world. In 2008, FAO estimated that rising prices plunged an additional 41 million people in Asia and the Pacific and 24 million in sub-Saharan Africa into hunger.

In these circumstances, the poorest, landless and female-headed households are always the hardest hit, and children, pregnant women and lactating mothers face the highest risk. Even in countries with a large proportion of people engaged in agriculture, most people buy food and are adversely affected by rising food prices. Poor people are disproportionately affected because they spend a larger share of their income on food. In trying to cope with the burden of consecutive food and economic crises, they cut expenditures on health and education or sell productive assets, creating poverty traps and negatively affecting longer-term food security. In Latin America and the Caribbean, livestock industries were disproportionally affected during the crisis by high fuel prices as transport and logistics costs are a high proportion of total production and marketing costs in this region. Fuel-importing countries were at a particular disadvantage.

Sources: FAO, 1998; Sponberg, 1999 ; CARE, 1998; USAID, 2009; FAO, 2008a; FAO, 2009a; World Bank, undated.

ing times of social upheaval (Ashley and Sandford, 2008). Diversifying livestock enterprises between small and large stock is a sound strategy for food security since small animals reproduce faster while large animals have greater value.

Keeping livestock also allows farmers to stabilize their income and consumption by selling eggs and milk on a regular basis and selling small animals such as poultry and guinea pigs at need. Dairy development projects that link smallholder farmers to markets promote food stability by securing regular income. Livestock help preserve and build the human capital that provides the family's active workforce by paying for medical bills and education; there are numerous reports of income from livestock contributing to these expenses (Nakiganda *et al.*, 2006; Rymer, 2006). They may also build social capital to help a family through a crisis. Smallholders and pastoralists will sometimes lend or give animals to relatives, knowing that this gives them social standing and puts them in a stronger position to ask for help in the face of a disaster. Because of their portability, livestock have a special role to play when people are physically displaced by conflicts or natural disasters. A family can move animals, but must leave buildings and crops behind.

Livestock owners respond in different ways to crises. In northern Kenya, pastoralists are reported to build their herds (particularly breeding animals) in times when feed is plentiful (Bailey *et al.*, 1999; Umar and Baulch, 2007) and sell them during droughts to cover essential expenses. In India, buffalo owners are reported to sell their animals to cover expenses (Rosenzweig and Wolpin, 1993). On the other hand, pastoralists in West Africa are reported to hold on to their animals even in times of food insecurity, possibly choosing not to sell large animals at a time when prices are low (Kazianga and Udry, 2006; Fafchamps *et al.*, 1998; Pavanello, 2010), preferring to retain them to start again when the crisis is over. They use other coping mechanisms such as skipping meals and increasing reliance on tea and sugar intake.

In systems where destocking and restocking

are normal practice, breeding females are maintained so that the herd can be rebuilt when conditions improve and only sold in extreme emergencies, but if a crisis becomes prolonged, animals of any age and sex may be sold. Small livestock are a convenient buffer against shocks for several reasons: they require lower capital investment, they are easier to sell quickly, if one dies it is less damaging, they grow and breed faster, and they survive on harsher terrain (Costales *et al.*, 2005). It is often the small livestock owned by women that are sold at short notice to cover periods of income deficit.

At global and national levels, the livestock sector can provide a buffering effect for food system stability. In a severe economic crisis, global consumption and production of meat falls, thus freeing cereal grains for other uses and damping down price shocks for staple foods (FAO, 2009b). Nationally, livestock production for domestic use can contribute to food security by buffering countries against problems with international food supplies. Livestock exports also have the potential to make an important contribution to the national balance of payments for countries that are net exporters.

International trade can make an important positive contribution to food security but it exposes countries to volatility in international markets. Additionally, export subsidies and tariff and non-tariff barriers of both developed and developing countries bring cheap, subsidized imports into developing country markets. It is said that small-scale livestock producers cannot match the higher quality and lower prices of imported products and are squeezed out of their traditional markets (Costales *et al.*, 2005). However, an economic analysis of milk powder imports in six countries found that, in many cases, milk powder was primarily sold in major cities, which means rural dairy producers selling milk in rural areas would not be affected by the competition (Knips, 2006). There appears to be limited evidence that dairy imports affect the welfare of most producers, market agents or consumers (Jabbar *et al.*, 2008). As for exports, in developing countries where not all livestock owners can take advantage of export markets, the poorest tend to benefit least. In the Horn of Africa, for instance, where livestock export has been growing, richer producers and traders have been able to benefit from the variety of export markets while some poorer herders have been forced by economic circumstances to sell animals and become contract herders (Aklilu and Catley, 2010).

VULNERABILITY TO CLIMATE CHANGE

While livestock contribute to food stability, livestock systems face threats to their own stability. One aspect of vulnerability is manifested in the effects of long-term trends associated with climate change, the increasing need to find renewable forms of energy and the growing human population displacing grazing livestock systems. Recurring droughts in the Horn of Africa have forced poor pastoralists and agro-pastoralists to sell animals that they might not normally choose to sell, to diversify their herds (Pavanello, 2010) and to rely on a wider income range than livestock ownership (Ashley and Sandford, 2008). In Burkina Faso, the successive droughts of the 1970s and 1980s led to a depletion of natural resources and migration which, accompanied by vague land tenure laws, became key constraints to livestock owners securing pasture and water (Gning, 2005). Livestock markets are one way to improve the ability of these producers to regulate stocking rates. Various government-regulated schemes have been tried in the past but today there is increasing focus on the functioning of private markets. However, lack of infrastructure, distance between producers and consumers, high transactions costs (Okike *et al.*, 2004) and poor price information are still constraints in many places. Well-designed restocking schemes (LEGS, 2009) can help livestock owners restock after a serious disaster when normal restocking mechanisms are overloaded

There are often links among access to grazing land, conflict and environmental degradation which can affect the food security of poor livestock owners. For example, tension exists between pastoralists and settled farmers in the Intergovernmental Authority for Development (IGAD) region that stretches across Djibouti, Eritrea, Ethiopia, Kenya, Somalia, Sudan and Uganda. Here, land tenure policies that have not clearly defined the rights of land users and have allowed privatization of grazing land for agricultural purposes are often at the centre of conflicts (Ashley and Sandford, 2008). Pastoralists who have lost grazing land to settled farmers suffer from restricted movement which leads to overgrazing and consequent environmental degradation. As a coping mechanism, some have chosen to keep smaller animals which they can sell quickly and use to buy cereals, or have reduced herd sizes to have more land for crop production.

DESTABILIZING EFFECT OF ANIMAL DISEASES

Occurrence of infectious animal diseases reduces the stability and resilience of the food supply from livestock, affecting everyone along the production and market chains. They can have four different effects: i) reducing the livestock population through death or culling; ii) reducing productivity of livestock; iii) creating market shocks when demand falls and supply contracts in response; and iv) disrupting international trade in livestock products. These effects can have impacts at macro and micro levels.

Rinderpest provides a dramatic example. Outbreaks in the 1890s killed approximately 80 percent of the cattle in southern Africa and caused widespread starvation in the Horn of Africa. One hundred years later, in the 1980s, the disease killed an estimated 100 million cattle in Africa and West Asia. A decades-long international control effort has resulted in the disappearance of clinical disease throughout the world. More recently, the global epidemic of highly pathogenic avian influenza (HPAI), which began in 2003–4, resulted in market shocks in a number of countries, the loss of 250–300 million poultry and the realignment of international trade

(McLeod, 2009). At the global level, the poultry sector recovered surprisingly quickly and widespread effects on food security were limited and short term. However, there were pockets of severe effects, such as in Cairo and Jakarta where family diets diminished when poultry were no longer available as a source of income and food (Geerlings *et al.*, 2007; ICASEPS, 2008). Other diseases have locally devastating effects such as *Peste des petits ruminants*, a disease that causes high mortality in sheep and goats, and has been reported several times in eastern and northern Africa since 2007.

Transboundary diseases place severe limits on international trade and have a high cost, but their precise effects on the stability of the food supply are hard to assess. FAO/OECD projections describe them as "damping down" export trade. For example, the bovine spongiform encephalopathy (BSE) outbreak in the UK in 1996 resulted in a 6 percent drop in beef consumption within the EU that took four years to return to previous levels (Morgan, undated). However, its impact on global consumption was obscured by a strong growth in demand from developing countries which compensated for reduced demand in the EU. There must have been some impact on meat supplies with culling of animals, but it was not reported. Similarly, the 2001 foot-and-mouth disease (FMD) outbreaks in the UK resulted in a large loss of animals through culling, including valuable breeding stock, but losses in supply from the UK were largely made up by supplies from elsewhere and there has been no estimate of the impact on the global food supply. When Brazil experienced FMD outbreaks in 2005, some parts of the country lost export markets but, by compensating internally, the industry as a whole maintained its export market share (FAO, 2006b).

The food supply also is impacted by a myriad of animal health problems that occur at community and herd levels. They decrease the productivity of animals by causing death or reducing the efficiency with which they convert feed into meat, milk and eggs (FAO, 2009b). These may cause chronic or seasonal losses and often require families to manage animals in risk-averse ways that reduce the level of production.

Poor livestock owners often face multiple shocks that hit at the same time, threatening their livelihoods and therefore access to food – a situation such as the sickness or death of an animal during a drought with prices of livestock feed increasing and prices for livestock products dropping. Crises can be recurrent or long term – in this respect, a livestock disease such as FMD that permanently reduces the productivity of an animal is a threat to resilience. For this reason, livestock's contribution to food security relies on a multi-faceted approach that builds resilience into the livestock sector and livestock-owning communities, and takes particular account of the needs of vulnerable people when planning and implementing crisis responses.

©FAO/F. McDougall

Access to food

Even when sufficient food is available within a country, households and individuals will only be food secure if they have the ability to access it. The majority of undernourished people suffer from lack of access, not from lack of food availability. Access requires people to have income to buy food or the means to barter for it. Food must be affordable within household budgets, and it must be available in convenient places and forms. There are social and cultural factors that affect entitlement to income and food, one of which is the gender dynamic within households and communities. Each of these elements of access is discussed in this chapter, which first examines the contribution that livestock make to accessing food of all kinds, and then reviews the affordability of and markets for foods of animal origin.

FINANCIAL, HUMAN AND SOCIAL CAPITAL

Livestock provide income and bartering power that contribute to their owners' ability to access food of all kinds. Livestock also contribute to

human capital and hence the ability to buy and produce food, by financing education and medical expenses. They can be a source of social capital, giving people a safety net to sustain them in food insecure times, through networks of gifts, loans and other transfers such as dowries. They provide income and employment not only to farmers but also to contract herders, animal handlers, traders, market operators and slaughterhouse owners and workers.

The contribution that livestock make to income is highly variable. A close look at 14 countries of the FAO Rural Income Generating Activities (RIGA) database found at least 50 percent of households in every country are recorded as keeping livestock and in some cases close to 90 percent. For those households, livestock are estimated to provide between 2 and 32 percent of their income (Table 8). The importance of livestock as an income source differs more by country than by income level.

There is no clear pattern of association between income levels and the contribution of livestock in Table 8 or in other sources. Several research reports link poverty levels and livestock ownership, but they use a variety of

TABLE 8

PERCENTAGE OF TOTAL INCOME OF RURAL HOUSEHOLDS COMING FROM LIVESTOCK ACTIVITIES, BY EXPENDITURE QUINTILES

COUNTRY AND YEAR	% OF HOUSEHOLDS OWNING LIVESTOCK	% OF HOUSEHOLD INCOME FROM LIVESTOCK IN EXPENDITURE QUINTILES					
		1	2	3	4	5	TOTAL
AFRICA							
Ghana 1998	50	20	19	19	17	16	18
Madagascar 1993	77	18	19	18	16	19	18
Malawi 2004	63	12	14	14	15	15	14
Nigeria 2004	46	6	5	5	5	5	5
ASIA							
Bangladesh 2000	62	6	6	8	8	7	7
Nepal 2003	88	18	22	23	24	26	23
Pakistan 2001	47	19	22	24	26	28	24
Viet Nam 1998	82	21	20	19	19	16	19
EASTERN EUROPE							
Albania 2005	84	32	29	23	25	20	26
Bulgaria 2001	72	7	16	17	17	15	15
LATIN AMERICA							
Ecuador 1995	84	15	16	17	18	15	16
Guatemala 2000	70	4	5	5	5	7	5
Nicaragua 2001	55	10	17	19	19	20	17
Panama 2003	61	2	3	6	5	7	5

Source: RIGA dataset, accessed September 2010.

variables, indicators, methodologies and data sources (Pozzi and Robinson, 2007). Although each contributes to understanding the role of livestock in household food security, they are hard to aggregate or compare. In a study of 16 countries, Delgado *et al.* (1999) found that the poorest households tend to be less dependent on livestock than those that are slightly less poor, while Quisumbing *et al.* (1995) found that poor households often earn a larger share of income from livestock than the wealthy. It is evident from the available information that livestock do contribute to the incomes of the poor, although perhaps less to the very poorest households who have no space to keep animals, cannot afford to feed them or find them too risky to own.

Livestock-owning households make choices about which of their animals or animal products they will produce to eat and which to sell, de-pending on their cash needs, access to markets and cultural preferences, but these do not fit into a universal pattern. In Bangladesh, for example, small-scale dairy farmers only consume a small amount of the milk they produce, selling most of it to meet immediate cash needs, even though milk is important to the Bangladeshi diet (Knips, 2006). Small-scale milk producers in Thailand, where milk is not a traditionally important part of the national diet, are responsible for almost all the country's milk production, but only consume 1 percent on-farm (Knips, 2006). In Cambodia, where meat is not central to the diet, livestock represent an important source of income rather than meeting immediate household food needs (Ear, 2005). A 2006 report from Senegal (Kazybayeva *et al.*, 2006) found relationships existing among geographic location, livestock type and the role of livestock in poverty alleviation in that country. In Viet Nam, rural poultry owners sell a smaller proportion of their product than those in peri-urban areas (Hancock, 2006). By contrast, a study from Nepal (Maltsoglolu and Taniguchi, 2004) found that livestock made a very important contribution to total household income in isolated hill and mountain areas with limited access to markets and cash income sources.

There are many potential uses for income from livestock (Nakiganda *et al.*, 2006). The proportion spent on food will depend on family needs at the time. A poultry project in Bangladesh resulted in asset accumulation from increased income, which was spent on education, improved housing, fencing, latrines, bedding, furniture, other livestock and investing in a family business (Dolberg, 2003). A more direct relationship can be seen in the IGAD region of East Africa where pastoralists and agro-pastoralists sell their high value livestock products and buy low cost cereal products for consumption (Ashley and Sandford, 2008). A community-level poverty assessment in three districts of Western Kenya (Krishna *et al.*, 2004) found that as households climbed out of poverty, they spent money on (in order of priority): food, clothing, shelter, primary education and then small animals, at which point they were no longer considered poor. At the same time, the loss of livestock can cause a household's descent into poverty, due to factors such animal disease, theft or an unplanned sale or slaughter to meet heavy funeral or human health expenses.

National livestock policies as well as national attitudes towards the role of livestock in agriculture have a significant impact on livestock production. By supporting or constraining the incomes of small-scale livestock producers, they also have indirect influence on access to food. Some national policies fail to promote livestock production or consumption in a way that favours the poor. Livestock are under-represented in most Poverty Reduction Strategy Papers (PRSPs), and even when they are considered, it tends to be in relation to the potential for boosting national GDP rather than alleviating poverty (Blench *et al.*, 2003). Such support tends to favour wealthier producers at the expense of poor producers, and focuses on livestock and technical issues rather than on people and poverty reduction (Ahuja *et al.*, 2009). This may be due to a misunderstanding among policy-makers who do not consider that livestock are a key income source for the poor or that pro-poor livestock production policies are important (Ashley and Sandford, 2008).

In addition, poorly planned attempts to reduce public spending through privatization of veterinary services have resulted in under-funded state veterinary and livestock extension systems and a private sector incapable of filling the gap, leaving small-scale livestock owners highly vulnerable to losses from epidemic and endemic diseases. The livestock producers who can organize themselves sufficiently to make demands on the government tend to be exclusive and not pro-poor. The fragility of the livelihoods of small-scale producers and pastoralists in countries such as Ethiopia, Senegal and Bolivia demonstrates the damage that such unsupportive policies can do to small-scale livestock production (Gning, 2005; Fairfield, 2004; Jabbar, *et al.*, 2008; Halderman, 2005; Ear, 2005).

Some plans and policies have been more supportive. For example, the Indian government's 11th 5-year plan pledged more equitable benefits from poultry production for small, marginal and landless farmers (Pica-Ciamarra and Otte, 2009). In Thailand, the recent rapid increase in milk production has been largely thanks to government support to cooperatives, credit access and training in the dairy subsector (Knips, 2006). The Thai government's support to the dairy subsector has been accompanied by a government school milk programme. In Kenya, positive dairy development policies once provided a regulatory framework, quality control, breeding services, animal health inputs, research, extension, pricing and tax policies, and expansion of rural infrastructure such as roads (Jabbar *et al.*, 2008). As a result of these policies, which were backed by the private sector, small-holder dairy farmers came to dominate production until the early 1980s. However, subsequent reduced budget allocations led to a decrease in the quality of services, and policies did not recognize the activities of the burgeoning number of farmers selling milk, milk bar operators and milk transporters, whose activities were effectively rendered illegal. In 2004, the dairy policy was revised to allow the Kenya Dairy Board to license and train small-scale traders (Kaitibie *et al.*, 2008).

Government policies also have directly promoted the food security of consumers through food assistance programmes. In Peru, for example, the government spends approximately US$200 million a year providing milk and milk products to the poor and children through food assistance programmes (Knips, 2006).

GENDER DIMENSIONS OF ACCESS

Gender dynamics are important in the food security of families and individuals, particularly the poor. They influence who can earn income or gain social capital from livestock, and the way that animals are managed which impacts how they contribute to animal source foods produced for the household. Gender dynamics also affect the way food is divided within families, especially in time of shortage. All of this can add up to greater or lesser food security for individuals and the family as a whole. Things play out differently across countries and social settings and the picture painted here is a broad-brush summary of what has been reported for developing countries.

Women contribute to producing income from livestock, alone and in partnership with male family members. Their ability to do this is constrained by limited access to inputs and services and by cultural norms that affect their daily lives. However, there is limited information on the way that gender dynamics change and roles that women play when livestock systems scale-up and concentrate beyond a certain level. Most of the information on gender influences on live-stock production, productivity and income are from research reports derived from studies of small-scale farms in rural areas of developing countries.

One way of considering the effect of gender is to compare male- and female-headed households. In 10 of the 14 countries shown in Table 8, livestock contribute a noticeably greater percentage to household income in male-headed than female-headed households (Table 9), especially in the African and Asian countries. In the Latin American countries, there is no difference or livestock make a greater contribution in female-headed households. Where there is a difference in income between male- and female-headed households, it is likely to be a result of difference in herd and flock sizes. Female-headed households have lower access to resources such as credit and labour, which restricts the number of animals they can own. However, with the animals they have, they are as productive as male-headed households (Pica-Ciamarra *et al.*, in preparation).

Whether heading a household or operating within a male-headed household, cultural biases in many countries constrain women's access to services of all kinds and this, along with their limited or nonexistent individual entitlements to natural resources, is associated with a lack of incentive to be more productive (Geerlings *et al.*, 2007; Quisumbing *et al.*, 2004). For example, there are numerous stories of women being excluded from animal production and health training because it is offered only to the heads of households, being unable to access credit because they have insufficient collateral, or not being directly informed about emergency animal disease control measures because the information is given out at a place or time that does not take account of their daily schedules.

Women are likely to own or have control over smaller livestock although they may have access to the products of larger livestock. The main exception to this is ownership of improved dairy animals, often provided through projects. Small livestock, as well as dairy products, are widely

TABLE 9

PERCENTAGE OF TOTAL INCOME COMING FROM LIVESTOCK ACTIVITIES, BY SEX OF HOUSEHOLD HEAD AND EXPENDITURE QUINTILE

	HOUSEHOLD HEAD	Q1	Q2	Q3	Q4	Q5
Ghana 1998	Female	14	12	12	11	11
	Male	22	23	23	19	18
	M/F	1.6	1.9	1.9	1.7	1.6
Madagascar 1993	Female	13	13	12	10	14
	Male	20	20	20	17	20
	M/F	1.5	1.5	1.7	1.7	1.4
Malawi 2004	Female	10	13	13	16	14
	Male	12	14	15	15	15
	M/F	1.2	1.1	1.2	0.9	1.1
Nigeria 2004	Female	3	2	3	4	5
	Male	6	5	5	5	5
	M/F	2.0	2.5	1.7	1.3	1.0
Bangladesh 2000	Female	3	3	4	3	4
	Male	6	6	8	9	7
	M/F	2.0	2.0	2.0	3.0	1.8
Nepal 2003	Female	10	19	16	18	18
	Male	19	22	23	23	24
	M/F	1.9	1.2	1.4	1.3	1.3
Pakistan 2001	Female	15	14	13	14	13
	Male	19	23	25	27	31
	M/F	1.3	1.6	1.9	1.9	2.4
Viet Nam 1998	Female	16	15	16	15	14
	Male	22	21	20	20	16
	M/F	1.4	1.4	1.3	1.3	1.1
Albania 2005	Female	19	22	17	20	6
	Male	32	29	24	25	22
	M/F	1.7	1.3	1.4	1.3	3.7
Bulgaria 2001	Female	8	5	12	11	14
	Male	6	19	19	20	15
	M/F	0.8	3.8	1.6	1.8	1.1
Ecuador 1995	Female	14	21	20	13	17
	Male	15	16	17	19	15
	M/F	1.1	0.8	0.9	1.5	0.9
Guatemala 2000	Female	7	6	4	6	7
	Male	4	5	6	5	7
	M/F	0.6	0.8	1.5	0.8	1.0
Nicaragua 2001	Female	8	12	16	13	14
	Male	11	18	20	21	22
	M/F	1.4	1.5	1.3	1.6	1.6
Panama 2003	Female	3	2	3	4	7
	Male	2	3	7	5	7
	M/F	0.7	1.5	2.3	1.3	1.0

Source: RIGA dataset. The figures used are the most recent available in the dataset for each country.

identified as resources over which women have both access and control.

There are significant examples of women earning income and contributing to food supplies by joining dairy producer cooperatives. In India and Pakistan, women are members of many of the cooperatives built around large specialized milk herds that meet urban milk demand. There are also a few reports of small-scale, independent individual producers who have invested in more intensive production units based on special milk breeds, improved feed regimes and improved disease control (Okali, 2009). There is no detailed demographic information about the women involved, except perhaps that they are poor.

There are two important points to consider. First, both women and men (husbands and wives) are involved, at times, in a joint activity. Apart from the cost of the animals themselves, these small-scale systems may use hired labour and depend on purchased feed, which suggests that only wealthier individuals can invest in these new intensive production systems. Second, the new institutional arrangements provided by the cooperatives have enabled poor women to overcome constraints to their access to services and credit (Arpi, 2006). The cooperative reduces the risk for actors at the lower end of the chain while enabling them to contribute to increasing the availability of livestock products through new markets. It also facilitates the investment required to ensure that food safety rules are followed.

Outside of programmes designed to ensure women's access to livestock, there is some evidence of women losing their access to milk animals at widowhood and divorce (Okali, 2009). Equally, there is evidence that individual women, and especially poor women, are not able to manage intensive systems on their own. Under these circumstances, the capital asset is likely to be viewed as a joint or household asset in which most members have some interest. Given that the animals are kept close to or even within the living quarters, this would not seem to be an unrealistic expectation (Okali, 2009). On the other hand, in a number of areas, especially in Southern Africa and Latin America, there is some suggestion that animals acquired by women through projects are treated differently, regardless of their size. In these cases, they are not socially embedded and control over the animals and the income gained from product sales is unlikely to be challenged.

In Bangladesh, BRAC's[1] poultry programme provides support for poor women and bypasses gender-biased public services. There is only limited information on the impact of these activities on livelihoods and food security, although there is some suggestion that the women involved in the BRAC poultry programme can climb the "livestock ladder" by acquiring a larger number of poultry and exchanging them for a more valuable animal.

The importance of poultry production for maintaining the nutritional well-being of poor households is emphasised in much of the livestock literature. In a number of countries, poultry production is presented as their main or even only source of protein, yet evidence from the H5N1 Highly Pathogenic Avian Influenza outbreak demonstrated that poultry are difficult for them to sustain in the face of disease outbreak and control measures. The food security impact of this is quite situation specific. A study carried out in the poorest governorates of Egypt (Geerlings *et al.*, 2007) found that poultry income was often the only contribution women made to household income, and if those contributions were reduced, their ability to negotiate with male relatives for money to fulfil their food security obligations was reduced, causing tension and intra-household conflicts.

In terms of decision-making over livestock sales and use within household flocks and herds, Nyungu and Sithole (1999) found that small livestock in both backyard and small-scale commercial systems must be seen as a joint house-

---

[1] BRAC, originally the Bangladesh Rehabilitation Assistance Committee but now known solely by its acronym, is a development organization based in Bangladesh, well known for its work with small-scale poultry producers.

hold resource, even if those individual animals were acquired by different individuals. As a joint resource, decisions about use, including sale, are likely to be open to negotiation, even joint decision-making, with final decisions dependent on need and who is present at the time. Small livestock may be seen simply as "small things", not to be bargained over, especially in circumstances where it is difficult to protect livestock health. In these situations, mortalities are likely to be high, and the number of animals can fluctuate dramatically over time. The wider gender literature shows that not all decisions about benefit allocations and even work are bargained over, as in the case of animals managed as household flocks or herds. Rather, they may be taken for granted and therefore unquestioned (Bourdieu, 1977), not even regarded as an imposition by those who ostensibly might lose out.

When it comes to food allocation within households, preference for one household member or another can be displayed through their being served more or higher quality food, which means they will have higher caloric intake, more variety and the possibility of greater nutrient density (Gittlesohn *et al.*, 1997). There is almost a universal expectation of food allocation bias against females of all ages, and against younger household members (Gittlesohn *et al.*, 1997). The bias against women is accentuated during food shortages (Agarwal, 1992a; 1992b). Household members considered to be at greatest risk of lasting damage from malnutrition are pregnant and lactating women and pre-school children (Lipton and Longhurst, 1989).

However, there is no substantive information on preferential allocations of meat and other livestock products within households. In some societies, pregnant or lactating women receive special nutritional treatment. In Egypt, for example, there is a tradition of giving eggs to women for some days after they give birth. Children generally appear to have claims over milk, but while some gender literature suggests that women will usually choose consumption over sale of milk and other products, there is

also information suggesting that both women and men might choose sale over consumption, and indeed this may be a rational decision. There are stories of children being denied eggs because it might encourage an appetite for expensive foods. From a very detailed study, Leonard (1991) concluded that the nutritional needs of younger household members were likely to be protected in situations where they contributed substantially to the household labour force. Jackson and Palmer-Jones (1999) made a similar case for adult men based on calculations that went beyond simply hours of work completed. Other literature suggests that women are denied meat or would not be allocated the "best" cuts, yet they are more often than not the food servers and presumably in some situations have a practical advantage over who is given what to eat.

There is also information describing how women might, in private, work around norms or customary practices that deny them certain foods. They manage to improve their own food intake by manipulating food portions, snacking frequently, increasing their consumption of palliative foods during the hungry season – sugar cane and palm wine that have high energy content, palm nuts that can be chewed for a long time, or possibly even dried meat – planting larger gardens for vegetables when pregnant, cheating on food taboos, and resorting to subterfuge to access desirable foods (Bentley *et al.*, 1999).

ECONOMIC FACTORS AFFECTING CHOICE OF LIVESTOCK SOURCE FOODS

Livestock source foods are a choice for many people in many societies, as well as a valuable source of nutrition. However, their place in the household diet depends not just on preference but also on their affordability. This is affected by household income levels and the proportion of household income allocated to different kinds of food, and by the price of livestock source foods compared to crop-based alternatives. Each of these factors will be discussed in turn.

INCOME

Global statistics show that livestock source foods are fairly income elastic. As income levels have risen and urbanization increased, diets have changed. Demand for livestock products has diversified, consumption of livestock products has increased, and wheat and vegetable oils have been substituted for traditional foods such as cassava, maize and lard. These effects have been observable in many parts of the developing world, in poorer countries as well as emerging economies. Figure 6 demonstrates the close relationship between GDP per person and meat consumption per person in six regions, using annual data over a 40-year period.

Several country studies further illustrate the above relationship. In China, a survey of long-term trends showed that the diets of both richer and poorer people became more fat rich over time – with consumption of more vegetable oil

for the poorer and of more livestock products for the richer (Guo *et al.*, 1999). A study in Uganda and Viet Nam (Maltsoglou, 2007) found that increased incomes were matched with increased consumption of livestock products. Knips' (2006) review of six countries – Jamaica, Peru, Senegal, Tanzania, Bangladesh, Thailand – found that with rising incomes, accompanied by urbanization and westernization of diets, there has been a demand for diversified dairy products, such as pasteurized milk, ice cream and chocolate. Increased incomes also result in increased health and nutrition awareness which in turn leads to increased demand for higher value, safer and higher quality products (Costales *et al.*, 2005).

Conversely, low incomes are a major constraint to consumption of livestock products, particularly in poor countries. In Senegal, a litre of fresh milk in the capital, Dakar, could cost up to half the daily wage of a worker while in

6 RELATIONSHIP BETWEEN GDP PER PERSON AND MEAT CONSUMPTION PER PERSON PER DAY IN SELECTED REGIONS

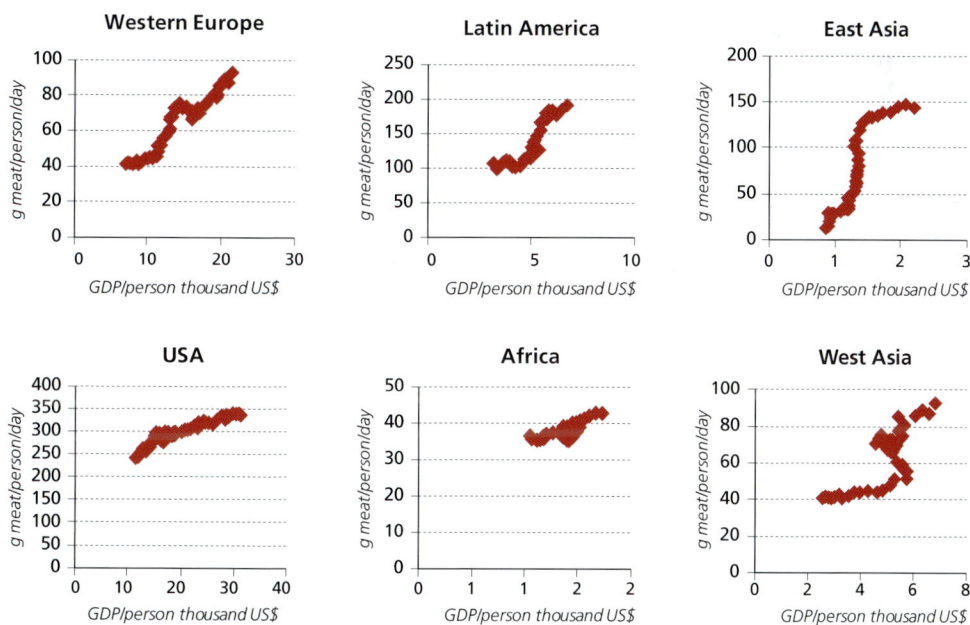

Source: http://www.ggdc.net/maddison/ and http://faostat.fao.org/site/291/default.aspx. Based on annual data from 1967 to 2007.

7 GDP PER PERSON IN CURRENT US$ BY COUNTRY INCOME GROUPING

US$

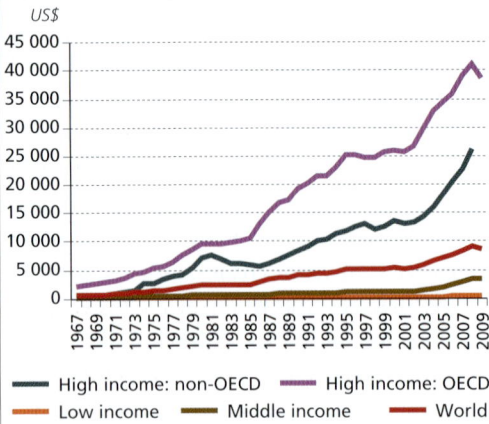

- High income: non-OECD
- High income: OECD
- Low income
- Middle income
- World

Source: World Development Indicators accessed January 2010.

St Louis Region, poor consumers could only afford to buy fermented milk and milk powder sold in individual servings (Knips, 2006). In Burkina Faso, most consumers prefer traditional poultry products and cannot afford the products from semi-intensive systems which are cheaper by the kilo but more expensive by the unit, the only measure poor consumers can deal in (Gning, 2005).

GDP per person, which is one measure of people's ability to spend, has been growing in most regions of the world. Between 1990 and 2008, it rose by 219 percent worldwide and by 207 percent in low income countries, although from a considerably lower base. Extreme poverty (people with an income of US$1.25 a day or less at 2005 prices) has been falling worldwide, from 1.9 billion people in 1981 to about 1.4 billion according to recent estimates. Overall this adds up to a slowly growing ability to purchase food, including livestock products.

Consumption of animal source foods is uneven across countries, regions and income levels, although the general trend is upwards. While developed countries have seen a slow growth in consumption from a very high base, the picture in the developing world has been more varied.

In East and Southeast Asia and particularly China, where economic growth and poverty reduction have been strongest, there has been a strong growth in consumption of livestock products. The countries within these regions that have higher per person incomes, such as Malaysia, Thailand and the Philippines, also have relatively high per person meat consumption (Costales, 2007). In China, GDP per person grew by over 1000 percent from 1990 to 2005. During the same period, the proportion of people living in extreme poverty fell from 60 to 16 percent. Consumption of meat rose from approximately 26 to 54 kg per person per year, milk from 7 to 26 kg, and eggs from 17 to 19 kg (FAOSTAT; WDI, 2010).

In South Asia, the poverty rate fell slightly from 1990 to 2005 but the number of people living in extreme poverty did not. While GDP growth in India was slightly above the world average, that of Bangladesh was below it. South Asia has seen a small rise in consumption of meat and eggs and a larger rise in milk consumption, with cultural factors playing a part (many Hindus are vegetarian), as well as a rise in small-scale dairying, making milk easily accessible to farm families.

In Latin America and the Caribbean, consumption of livestock products tends to be higher than in other developing areas and has increased rapidly. Countries such as Chile, Brazil and Ecuador have seen growth in GDP and a drop in poverty rates along with a large increase in consumption of livestock products while growth in other countries has been slower.

In Africa, there has been some growth but from a very low base. In many sub-Saharan African countries, GDP growth was up to 150 percent between 1990 and 2005 and, while the poverty rate for the region fell from 58 percent in 1990 to 51 percent in 2005 (calculations from Povcal), it is still high. The consumption of livestock products in the region also remained more or less static, with a slight decline in meat consumption and a slight rise in milk consumption between 1992 and 2002 (Rae and Nayga, 2010).

In addition to regional differences, there are differences between urban and rural consumption. In both poor and emerging economies, urban dwellers tend to have higher incomes and buy more livestock products through formal channels, particularly higher value processed products. A review by Maltsoglou (2007) reported that in Peru, Uganda and Viet Nam, urban households consumed 1.5 to 2.5 times as much livestock source food as those in rural areas. In India, urban consumers eat 2.8 and 4.5 times more eggs and poultry meat respectively than rural consumers (Mehta *et al.*, 2003) while in China, urban people have three times the income of those in rural areas and consume four times as much milk and twice as many eggs (Ke, 2010). In Thailand, 95 percent of dairy products are sold to urban consumers (Knips, 2006). Rising urban income in Bangladesh has led to a rapidly increasing urban demand for milk products, including pasteurized milk, milk powder, flavoured milk, sweet curd, sweet meats, ice cream, ice lollies and chocolate (Knips, 2006).

PRICE

Livestock source foods are rarely listed among household staples. They are more expensive than the grains and starches that provide the basic energy supply and often more expensive than plant-source protein such as lentils or beans. High prices depress consumption levels of livestock products. In Jamaica, for example, high production costs for fresh milk have led to decreased demand because consumers cannot afford it (Knips, 2006).

Worldwide prices of food in general, including livestock source foods, were about 40 percent lower in the mid-1990s and early-2000s than they are today and a little more stable (IMF, undated). In recent years, increasing grain prices have had a double impact on livestock – they have raised the price of staple cereals, reducing people's purchasing power and, at the same time, raised the cost of livestock feed. Interestingly, during the 2007–08 global economic crisis, meat prices increased less than cereal or dairy

product prices, but still the growth of demand for livestock products slowed. In richer countries, such as UK, this manifested as a change to cheaper cuts of meat, affecting people's lifestyle but not their food security. In poorer countries, there has been some substitution of crops for livestock protein.

Fish are also an important protein source and farmed fish, being efficient converters of feed, are a growing competitor to livestock. Maltsoglou (2007) found that in Uganda, poor families eat more fish than meat, while richer families eat more meat than fish. In Viet Nam, families in all wealth categories eat more fish than meat, while in Peru, meat is greatly preferred to fish, regardless of wealth status.

It is challenging to balance the need of producers to make a living with consumers' need for affordable food. In Viet Nam, for example, supportive government policies to develop the domestic dairy subsector have resulted in high milk yields, better dairy genetics, better dairy management and a rapid growth in production. However, the strong profitability of milk producers relies on substantial government support which maintains high output prices and low input prices to the disadvantage of the poor consumer – Vietnamese consumers pay European prices for milk (Garcia *et al.*, 2006). One reason for the rapid rise in chicken consumption has almost certainly been the fact that chicken meat it is relatively cheap compared to other meats (FAO, 2007).

MARKET ACCESS AND FOOD ACCESS

Access to livestock source foods is facilitated by the connections that producers and consumers have to markets for livestock products, which range from selling to one's neighbour over the fence to supplying supermarkets in distant cities through integrated market chains. Good market access increases the food security of producers through assured income and the food security of consumers by ensuring that food products will be locally available when needed.

Small-scale producers, pastoralists and poor

consumers do the bulk of their trading through informal markets and often close to home. Formal markets are almost non-existent in remote areas, and rural livestock producers face long distances, poor road networks and high transactions costs (Costales *et al.*, 2005). These factors encourage producers to consume at home and sell milk, meat and eggs in local marketplaces. Closer to town, peri-urban livestock producers have the advantage of proximity to a wider range of markets, so the prices they fetch for their produce are higher. They also benefit from increasing demand for livestock products, a result of rising incomes in urban areas. However, they still face barriers to entering formal markets due to requirements to meet consistent quality standards and volume, and for certification of product safety.

Much recent literature on livestock development and a great many development projects are concerned with linking small-scale producers to larger or more formal markets. The assumptions behind these efforts are that small-scale producers will have more lucrative and stable livelihoods if they are more strongly connected to semi-formal or formal markets, and that this will provide an incentive for them to become more efficient and productive. There is also, sometimes, an assumption that formal markets will ensure safer food for consumers.

DIVERSE MARKETS FOR DAIRY PRODUCTS

The greatest potential for connecting small-scale producers and traders with markets probably lies with dairy products, although not at the same level in every region. For example, Brazil, Latin America's largest dairy market, has intensified production considerably which implies limited prospects for small producers (Bennett *et al.*, 2006). However, in peri-urban areas of South Asia and some parts of Africa, there have been successful efforts to build market chains based on smallholders (Box 6). Dairy production benefits less from economies of scale than other livestock enterprises and provides a fre-

quent and regular income for those who produce and sell milk. The large size of the informal market, probably around 80 percent of marketed milk in developing countries, means that there is still scope for smallholder engagement. The perishable nature of fresh milk also lends itself to marketing close to where it is produced. For these reasons, small-scale peri-urban dairy marketing systems have the potential to make a growing contribution to food production in some regions, at the same time allowing consumers a choice as to where they buy their dairy products.

CONCENTRATION OF POULTRY MARKET CHAINS

Poultry systems are a complete contrast to dairy systems. Poultry production and marketing benefit from economies of scale. They exhibit distinct differences between the very large companies that dominate worldwide supply and trade and the small-scale producers in developing countries. As a country's economy grows, the informal peri-urban market initially thrives as entrepreneurs take advantage of new demand, but soon the subsector intensifies and small-scale producers and traders cannot compete. Concerns about hygiene also encourage urban councils to replace live bird markets with slaughterhouses that charge fees for processing. All of these factors mean that projects to connect small-scale poultry keepers to formal markets face a number of challenges and potentially a short life.

The few successes for commercial poultry smallholders have mainly been in local specialty markets. In Viet Nam, small- and medium-scale duck breeders and traders still predominate, supported by strong demand and little competition from industry. Recent projects to promote biosecure traditional chicken keeping in Viet Nam are also showing promise (Ifft *et al.*, 2007; USAID, 2007). In India, the KeggFarm poultry breeding company has produced a crossbred chicken that has meat similar to a traditional bird's but is suitable for outdoor living. The company has set up

---

BOX 6

INFORMAL MARKETING OF DAIRY PRODUCTS IN SOUTH ASIA, EAST AND WEST AFRICA

Successful small-scale milk marketing initiatives have added structure to informal dairy markets without excluding small-scale operators.

South Asia

In India, approximately 50 percent of milk is consumed by the people who produce it. Of the milk sold, 80 percent or more passes through informal channels – in 2002, an estimated 80 percent of Indian towns received milk only through informal markets (CALPI, undated). The milkman is often the only means by which the producer can sell and the consumer can buy milk on a daily basis. The well known "Operation Flood" project, which introduced more formality to milk marketing chains, was designed to meet the needs of small-scale operators with frequent local collection and regular payments.

In Bangladesh, 97 percent of milk is sold to milkmen, who then either sell it as sweets to sweet shops or to the consumer as fresh milk, curd or butter oil (Garcia *et al.*, 2004a).

East and West Africa

In East Africa, an estimated 80 percent of milk is sold through informal channels but the milk market varies by country.

In Kenya, dairy products are the largest item of food expenditure (Argwings-Kodhek *et al.*, 2005; Salasya *et al.*, 2006). More than 85 percent of milk is marketed through informal channels, which provide producers with higher prices than formal channels (Omore, 2004). A dairy policy passed in 2004 allows for the licensing and training of small-scale traders (Kaitibie *et al.*, 2008), meaning they participate in the market legally and can build more stable businesses.

In Tanzania, 90 percent or more of milk is consumed on-farm or sold to consumers close by, due to the inaccessibility of markets. In those parts of the country where cattle are not kept, milk consumption is very low (Knips, 2006). In Ethiopia, an average of 76 percent of all domestic milk production is consumed on-farm (Jabbar *et al.*, 2010).

In West Africa's Sahel countries (Kamuanga *et al.*, 2008), poor roads and lack of refrigerated trucks result in high transport costs and therefore low profits for rural producers. Even when they do manage to reach the markets, they have to sell door-to-door or from kiosks in the suburbs. Consequently, 80 percent of milk produced in rural areas of Senegal is consumed on-farm (Knips, 2006).

---

a marketing chain involving hundreds of traders with bicycles to supply fertilized eggs and chicks to village producers (Ahuja *et al.*, 2009). Once the birds are grown, producers find a strong demand in their local markets.

In spite of the dominance of large producers, village poultry consumed at home or sold locally are still important to food access in rural economies and likely to persist. Reports on poultry keeping in Africa frequently mention the importance of village chickens in providing meat and eggs for home consumption, often

indicating that around 50 percent of production is consumed at home. In Viet Nam, poor households that own small numbers of poultry as scavenging flocks use them mostly for home consumption (Maltsoglou and Rapsomanikis, 2005). The proportion of poultry consumed and used for other purposes within the household is much greater in the highland areas than in lowland areas which have better access to markets (Tung, 2005). In Bangladesh, the landless poor have a great need for income and thus are more likely to sell their poultry than consume them.

©FAO/Giulio Napolitano

MARKETING OF LIVE ANIMALS FROM PASTORALIST SYSTEMS

For pastoralists, the key to food access is a sustainable livelihood from marketing live cattle and small ruminants, often across international borders. Ethiopia is estimated to have exported 297 600 animals in 2007–08 with a value of US$41 million (Aklilu and Catley, 2009). A dependable and flexible market allows producers to regulate stocking rates and earn income. However, in the Horn of Africa, market access is affected by wealth, with better-off pastoralists having access to more markets, as well as by mobility, the types of animals owned and the pastoralist's position in social networks. While well-off pastoralists have benefitted from a growing export trade, those less well-off have suffered, with many losing their animals and becoming contract herders (Aklilu and Catley, 2009). Livestock disease also limits the markets that pastoralists can attempt to access. For example, the EU and USA markets are closed when diseases such as contagious bovine pleuropenemonia (CBPP) and FMD are present. They can also close existing markets suddenly. Outbreaks of RVF cut off exports from the Horn of Africa to parts of the Middle East in 1998, 2000 and 2007. Some traders were able to avoid the bans but other traders and livestock owners suffered badly from low prices and the inability to sell animals.

PROCESSING TO ADD VALUE AND PRESERVE PRODUCTS

Home processing of products is one way that rural livestock owners address market inaccessibility. In rural Bangladesh, farmers who have no access to markets for their produce process their milk within their households into traditional products such as ghee, channa and yoghurt, which can be consumed at home or sold or bartered in the village to rural consumers who have no access to high value dairy products such as pasteurized milk (Knips, 2006). In Peru, milk producers who are not located along formal milk collection routes usually convert their own milk into curd and sell it to local cheese-makers, who play an important role in maintaining dairy production in poor remote areas of the country.

Access to processing equipment can extend the shelf life of livestock products, but since much of this equipment is expensive, requires capital input and is subject to economies of scale, it is not an option for poor small-scale livestock producers (Costales *et al.*, 2005). The growth of processing centres in an otherwise remote rural region can remedy this situation and boost availability of livestock products. In Pakistan, one reason for increased domestic milk production is the presence of processing centres.

CONSUMER PREFERENCE

Both informal and formal markets are important to consumers. They generally prefer the taste and texture of meat from indigenous and extensively raised animals, and will choose them for holiday and special occasions. At the same time, they appreciate the lower cost of some products from intensive systems.

For rural consumers in developing countries, local markets may be the only ones within convenient reach. They provide lower prices, traditionally raised livestock and the opportunity to check the quality of products close to source.

In urban areas of developing countries, markets that offer fresh produce appeal to consumers who prefer to buy live animals and have them slaughtered at the market, rather than trusting the hygiene of food chains supplying meat. This has resulted in a proliferation of live-animal markets near or within cities. As will be discussed later under "City Populations", city councils have concerns about environmental and human health problems associated with these markets and would prefer them not to be there.

Supermarkets have taken over food supply in developed countries and are increasing their reach in the cities of the developing world (Reardon *et al.*, 2010). They offer the convenience of having everything under one roof, a consistent level of safety and quality and, for wealthier consumers, competitive prices. Integrated market chains that supply supermarkets are also easier to regulate in countries where a regulatory system and laws exist. However, for the many people who currently lack access to food, informal markets will continue to be important and so will "street food" bought in small quantities from stalls.

©FAO/PPLPI

Key points on livestock and global food security

Livestock make a necessary and important contribution to global calorie and protein supplies, but at the same time, they need to be managed carefully to maximize their contribution.

While livestock products are not absolutely essential to human diets, they are desirable and desired. Meat, milk and eggs in appropriate amounts are valuable sources of complete and easily digestible protein and essential micronutrients. Overconsumption, however, results in health problems.

Livestock can increase the world's edible protein balance by converting protein found in forage that is inedible to humans into forms digestible by humans. They can also reduce the edible protein balance by consuming protein that is edible by humans, from cereal grains and soya, and converting it into small amounts of animal protein. Choice of production systems and good management are important factors in optimizing protein output from livestock.

Livestock production and marketing can help stabilize the food supply, acting as a buffer to economic shocks and natural disasters for individuals and communities. However, the food supply from livestock can be destabilized, particularly by diseases.

Access to livestock source food is affected by income and social customs. Access to livestock as a source of income and hence food is also unequal. Gender dynamics play a part, particularly in pastoralist and small-scale farming communities, where female-headed households tend to have lower resources hence fewer, smaller livestock, and within families where the larger and more commercial livestock are often controlled by men. These problems are not unique to livestock, but they are prevalent among producers and consumers of livestock products and need attention.

The following section looks at three unique types of populations in terms of their relationship to livestock and livestock products: livestock dependent societies, small-scale mixed farming societies and urban dwellers.

Three human populations – three food security situations

©FAO/Vasily Maximov

Livestock-dependent societies

In societies that depend on livestock as their most important source of livelihood and food security, management of livestock shapes their way of life. These livestock-dependent societies have production systems based on grazing land. According to one definition (Sere and Steinfeld, 1996) at least 90 percent of the total value of farm production comes from livestock and more than 90 percent of dry matter fed to animals comes from rangelands, pastures and annual forages.

The largest number of livestock dependent people, currently around 120 million (Raas, 2006, based on data from 2002), is found in pastoralist societies, where livestock provide milk and occasionally blood and meat for their owners, carry the possessions of nomadic families when they move, are the main or only source of income when they or their products are sold, and the main capital asset owned by the family. Some communities practice mobile grazing,

moving animals over wide communal grazing areas, while others are sedentary graziers on communal grasslands.

Ranchers who keep animals extensively on the rangelands are another example of a livestock-dependent society, considerably fewer in number than pastoralists but important in their contribution to the total supply of livestock in their countries and the world. Animals are kept primarily for income, although they also make a direct contribution by supplying milk and meat to ranch families and employees. Ranchers and stock farmers often use grassland that they own or where they can have control over its use.

By definition, a livestock-dependent society relies heavily on livestock for its food security and livelihood and, as the discussion in this chapter will show, these societies have a special niche in global food security. At the same time, they face many challenges and need support in order to continue to play their important role. The level of production from these systems may be close to the limit, given limitations on natural resources, and they will increasingly need to rely on activities outside of agriculture for sustainable livelihoods.

CONTRIBUTIONS AND CHALLENGES TO FOOD SECURITY

Pastoralism and ranching contribute to food security in three important ways: they add to the total food supply, they strongly support food access by livestock owners and managers and, when managed appropriately, they contribute to a positive protein balance.

Rainfed grazing systems provide around 19.2 million tonnes of ruminant meat or 19 percent of world production (based on data in Table 6). They also provide about 12 percent of the world's milk. Ranching systems, which produce almost entirely for income, have more reliable access than pastoralists to higher value markets, putting them in a stronger position to contribute to global supplies. The Australian rangeland systems, for example, are the second largest producers of sheep meat in the world and export approximately 45 percent of their production (ABARE, 2010; Meat and Livestock Australia, 2011).

In some countries, pastoralism makes an important contribution to national food production and to GDP and, in a few cases, it also contributes significantly to export. The livestock of Mongolia produce one-third of the country's GDP and up to 21 percent of its export earnings. The rangelands of Morocco contribute an estimated 25 percent to agricultural GDP. Approximately 46 percent of the bovine meat in East Africa and just over 40 percent of small ruminant meat are estimated to come from pastoral systems (Raas, 2006) while in West Africa, pastoralism contributes 37 percent of bovine meat and 33 percent of small ruminant meat (Raas, 2006).

Livestock also have a very important function in supporting food access for pastoralist families. Their value is illustrated by the fact that, throughout the Horn of Africa, pastoralists define their wealth and poverty in terms of their ownership of livestock (Aklilu and Catley, 2009). In pastoralist households, all of the livestock source food may be produced from their own animals, and income from livestock makes up a large part of total household income. In Kenya, for example, livestock production is estimated to contribute between 50 and 95 percent of the income of pastoralist families (Aklilu and Catley, 2009; Kenya Ministry of Agriculture, 2008) while in Senegal, 80 percent of milk produced by pastoralist and agro-pastoralists is consumed by the household (Knips, 2006). Animals are also sold as needed to stabilize income or consumption during drought, or preserved to allow families to recover from disaster (Bailey *et al.*, 1999; Umar and Baulch, 2007; Pavanello, 2010).

Productivity from extensive grazing systems is low in terms of output per animal and per labour unit but high in terms of output from limited resources (water and grain). In these systems, livestock can be favourable to the protein balance because they use forage resources that cannot be used for any other form of food production. They also occupy land areas where there are limited alternatives for other types of production because good soil and water are in short supply, the terrain is hilly or the location is remote. However, dependence on livestock presents risks, because it occurs in fragile and challenging ecologies where there are limited prospects for diversification. Livestock owners are expert, specialized and their way of life is adapted to a harsh environment. They can be quite self-sufficient, requiring only limited inputs from outside. At the same time, the foundation of their livelihoods and food security, the livestock herd, is susceptible to disease, drought and harsh climate, and the output of individual animals is, on average, low.

Ranchers, who rely on selling animals or wool, have seen slower growth in demand for ruminant meat than meat from pigs and poultry. Production of ruminant meat has approximately doubled over the past 40 years, but has increased seven-fold for poultry meat (as shown in Table 4). Trade growth for beef and ruminant meat is also lagging behind the total for meat (Morgan and Tallard, undated). The trade shocks caused by diseases such as FMD and BSE hurt some countries that suffered outbreaks but benefited those that did not (Morgan and Tallard, undated). Climate change and difficult market conditions have caused ranchers in the USA, Australia and New Zealand to reduce their herds. The Australian national sheep flock has approximately halved in the past 20 years, even though there is growing demand for sheep meat in the Middle East. Ranchers cope with adversity by diversifying species and products, and investing in enterprises outside of livestock.

In pastoralist societies, people tend to be poor and often their livelihoods and food security are fragile. Forage and water are limited, theft of animals is common and disease outbreaks at times cause heavy losses to pastoralist herds. In recognizing that these are part of the normal course of events, management focuses on building resilience into the system, targeting stability rather than high levels of production (FAO, 2003; Mamo, 2007; Barrow *et al.*, 2007).

Mating in some systems is restricted to narrow windows of time to allow lactating animals to make best use of forage and allow young animals to grow during the most favourable weather. Destocking and restocking are used to cope with fluctuations in the forage supply, young animals being sold and the breeding herd maintained. Movements are timed to reduce exposure to raiders, and armed young men guard the animals. A number of measures are used to restrict exposure to disease and risks are weighed carefully. Quarantining of new animals, avoiding neighbouring herds when a disease outbreak has occurred in the vicinity, avoiding wildlife, controlling ticks and tsetse flies, and the use of antibiotics to cure CBPP are all risk management practices used by pastoralists.

There is limited potential to diversify livelihoods out of livestock other than by sending family members away to seek education and work in cities and foreign countries, which has the risk that they will not return. Loss of land through encroachment by settled farmers, development of wildlife areas or building of dams, as well as the threats of drought, conflict and insecurity are all identified as reasons why African pastoralists have migrated to urban areas. They

seek work in the informal sector, yet their livelihoods and food security do not necessarily improve (UN HABITAT, 2010). Others have acquired rights to land and become mixed farmers.

It is rare that an entire country represents a case study for one type of society or production system. Mongolia represents that unusual situation because to a large degree, the country as a whole could be said to be livestock dependent. The following case study examines the extent to which this is true and the way conditions are changing.

CASE STUDY
MONGOLIA: THE LIMITS OF THE LAST PLACE ON EARTH[2]

Mongolia is sometimes (and with respect) called "the last place on earth" referring to its remoteness and its open spaces. The popular image of Mongolia is of wide open steppe or desert dotted with white round tents (*gers*, sometimes called *yurts*) and nomadic herders on horseback following their flocks of sheep, goats, horses, cattle and Bactrian camels against a backdrop of mountains and a deep blue sky. A land with no fences, it is almost three times the size of France but has only 2.7 million people.

With livestock numbers at record high levels, solar charging panels on the roofs of many *gers*, and a satellite dish providing television reception in every community, livestock producers seem to be doing well. In the capital, Ulaanbaatar, shops are well stocked with televisions, computers and luxury consumer goods, a great change since the mid-1990s. Appearances are partly true, but they are no longer typical and serious problems often go unnoticed.

Mongolia is one of the last countries where livestock raising provides the greatest source of employment – around 40 percent of the population – and where few other forms of land use are possible. It is perhaps as livestock dependent a country as can still be found.

Mongolia is entirely landlocked, sandwiched

[2] The case study is adapted from Honhold, 2010.

8 HUMAN POPULATION OF MONGOLIA 1980 TO 2007

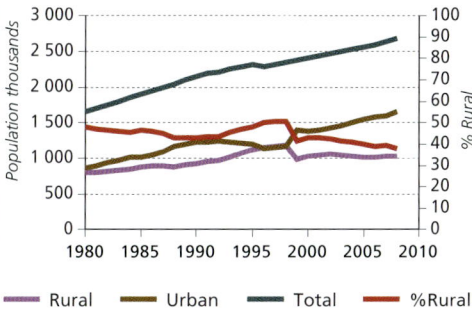

Source: Annual yearbooks of the National Statistical Office of Mongolia.

9 LIVESTOCK POPULATION OF MONGOLIA BY SPECIES 1980 TO 2009

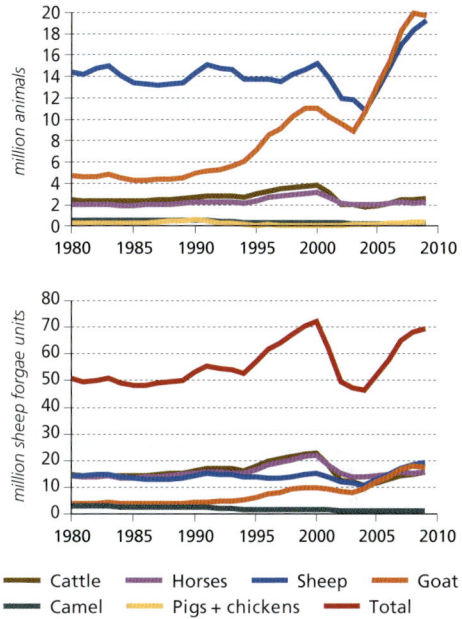

Source: Annual yearbooks of the National Statistical Office of Mongolia.

between Russia (Siberia) to the north and China (largely Inner Mongolia) to the south. The countryside is open, with virtually no fences, and ranges from desert to mountains to steppe to forest. However, where visitors see apparently empty, wide open spaces, herders see the countryside defined by water sources and wintering sites. These are limited and their number, particularly of the latter, is difficult to change. Water resources have been increased in the past by establishing wells, but this can have the effect of enabling livestock to use pastures that would otherwise be kept for winter pastures or fodder.

The current human population of Mongolia is around 2.7 million, with a population density of 1.7 per km², making it one of the most sparsely populated countries in the world. However, since 1977, 50 percent or more of the population has lived in urban centres, either the capital city or the major province centres. Figure 8 shows the growth in the total population since 1980 and the increasing proportion made up by the urban population.

The livestock population almost doubled between 1988 and 2009, increasing to around 44 million, almost entirely ruminant livestock and horses. There are very few chickens or pigs in the country (see Figure 9). However, this total

does not take into account the change in the composition of the national herd, in which the sheep and goat populations have grown rapidly in recent years. The Mongolians use their own measure of a livestock unit, a Sheep Forage Unit (SFU), to create equivalence between different grazing species in terms of the amount of forage each requires. Calculations using these units indicate the change in the size and composition of the national herd in relation to its use of forage, as shown in Figure 9. The total national herd size has risen from 50 to 70 million SFUs. The total was fairly level until around 1990, but since then, there have been rapid rises and equally rapid declines. These latter have been linked to the occurrence of severe winter conditions (*dzuds*) and summer droughts. As the rural population is dependent on livestock rearing for income, such fluctuations create obvious impacts on their livelihoods. Equally, the prices they obtain

10 MILK AND MEAT PRODUCTION IN MONGOLIA 1980 TO 2005

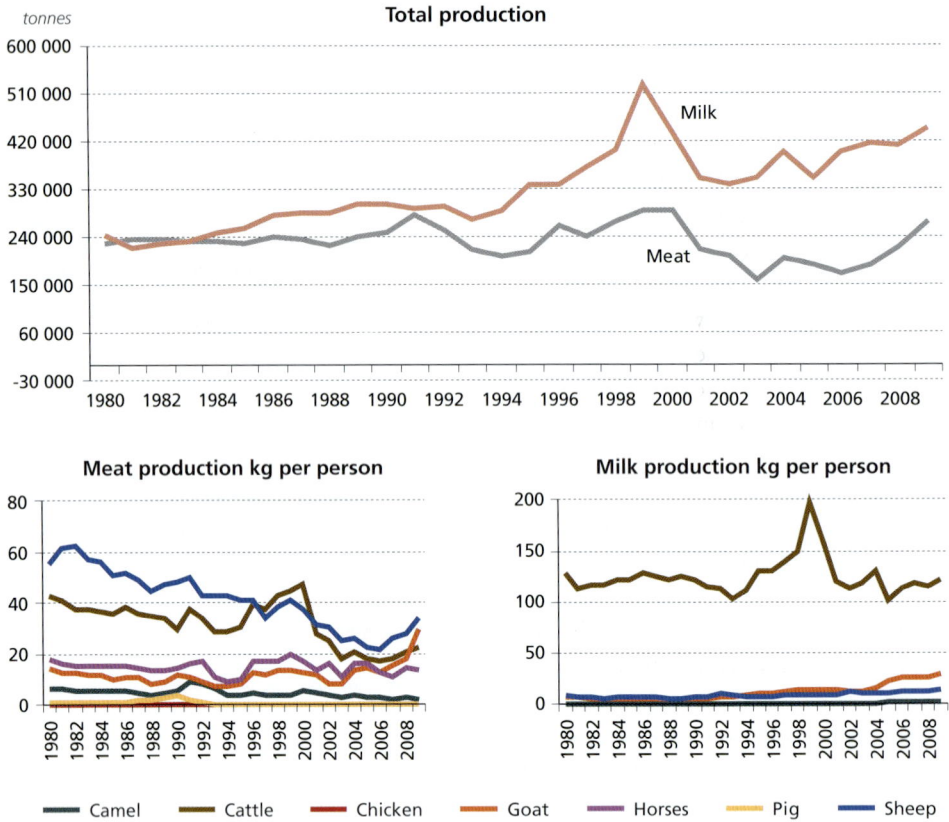

tonnes

Total production

Milk

Meat

Meat production kg per person

Milk production kg per person

Camel — Cattle — Chicken — Goat — Horses — Pig — Sheep

Source: FAOSTAT (production data) and World Development Indicators (human population).

for their products have a strong impact, and they have been affected by recent changes in the price of cashmere. The number of animals is a poor guide to the health of the livestock industry.

Livestock supply meat, milk, fibre and transport, although the last is decreasing. Meat production grew from 1961 to 1978, from around 150 000 tonnes to 230 000 tonnes but then levelled off until the late 1980s. Since then, as shown in Figure 10, total meat production has fluctuated from 280 000 to 150 000 tonnes annually and the species contributing to meat production have varied from year to year. Part of this fluctuation was due to a series of *dzuds* and droughts that occurred from 1999 to 2002. On a

per person basis, there was an overall downward trend in production between 1980 and 2009, despite record high numbers of livestock being kept. Many herders have turned to producing and selling cashmere as a cash crop, which is seen in the increased numbers of goats being kept. There is no reliable public database of cashmere production, therefore the figures are not reproduced here, but estimates in the early 1990s suggested that world production was around 4 500–5 000 tonnes per year, of which Mongolia supplied 20–25 percent (Petrie, 1995). Mongolian cashmere is generally of high quality and commands a good price for the raw product (de Weijer, undated), but this is a non-essential commodity largely supplying a luxury market where prices fluctuate (Schneider Group, undated).

FOOD SUPPLY

The daily dietary energy requirement recommended for Mongolia is 1 840 kcal per person per day (FAOSTAT, accessed October 2010).

Figures 11 and 12 show the average caloric intake per person per day in Mongolia between 1980 and 2007. Apart from a short period during 1991 to 1994, food supply was over 2 000 kcal per person per day, and the most recent trend was a gradual rise. However, the contribution from animal products declined over this period from just under 1 000 kcal to around 750 kcal per day, from 40 to 30 percent. The decline in the contribution of meat was more marked, with much of the difference made up by an increase in milk supply. Much of what is consumed is produced in Mongolia, including starchy roots (potatoes) and cereals (mostly wheat).

The makeup of the daily energy supply, even in nomad families, has always contained a significant contribution from vegetable products, particularly cereals. However, in the early 1960s, locally produced animal products contributed over 50 percent of a daily energy supply per person of just over 2 000 kcal. By 2007, this had fallen to around 33 percent of a daily per person supply of 2 300. During that time, the proportion of energy supply produced locally had fallen from around 90 percent to 50 percent. Per person meat supply has not kept pace with the increase in population and is falling despite increased herd size.

Sugars, vegetable oils, other cereals and other vegetables and fruits are mostly imported. The proportion of the calorific intake that is imported has risen from around 20 percent to 50 percent, largely from an increased import of cereals, although vegetable oils are increasingly important in the diet.

Cereal production was established in the 1960s through the virgin lands system, with monoculture systems using large fields and high levels of

11 KILOCALORIE CONSUMPTION PER DAY IN MONGOLIA BY SOURCE 1980 TO 2007

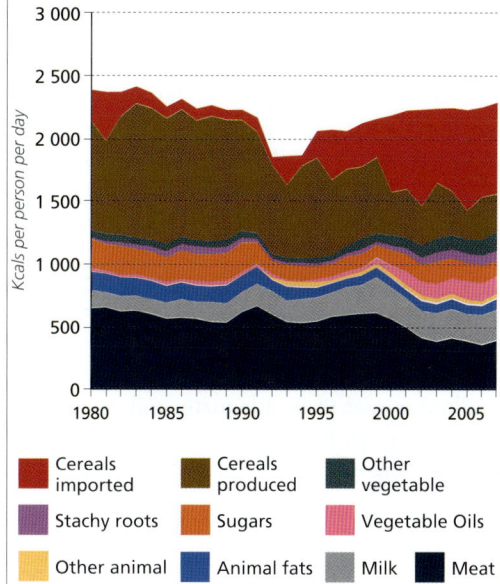

Source: FAOSTAT.

Legend:
- Cereals imported
- Cereals produced
- Other vegetable
- Stachy roots
- Sugars
- Vegetable Oils
- Other animal
- Animal fats
- Milk
- Meat

12 PERCENTAGE OF MONGOLIAN DAILY KILOCALORIE INTAKE THAT IS IMPORTED

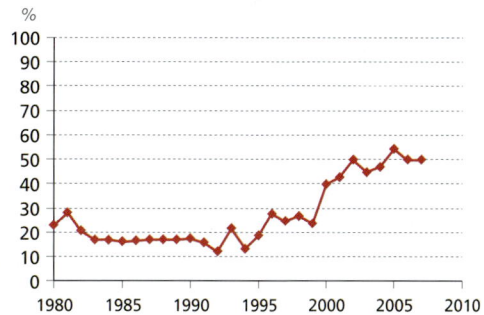

Source: FAOSTAT.

mechanization and irrigation established by the state. In the early 1990s, state support for these systems was withdrawn resulting in dramatic declines in locally grown cereals. The decline continued until 2008–09, when local wheat production increased again owing to government investment, and in 2009, Mongolia may have become almost self-sufficient in grains (contrary to Figures 11 and 12). However, the production system gives yields of 0.8 to 1 tonne per hectare (National Statistical Office of Mongolia, 2007), around 10 percent of that for farms in Europe and North America, and relies completely on imported fertilizers, fuel and machinery. National food security will be improved, but at a high financial cost.

FOOD SECURITY FOR HERDERS AND URBAN DWELLERS

Food security for herders (livestock dependent families) is said to be adequate, and on average this is likely to be true. At the time of change from largely government-owned livestock to private ownership in 1990, around 261 000 (58 percent) of 450 000 households, had some livestock. By 2007, this had fallen to 226 000 (35 percent) of 646 000 households (National Statistical Office of Mongolia, 1980 to 2009). The number of herds rose between 1990 and 1995, but has been declining since then. At the same time, the size of herds has fluctuated.

There is a minimum herd size to enable a herd to survive and recover from adverse climate events such as drought or *dzud*, below which the herder is considered poor and vulnerable. Different publications set levels for viability between 50 and 200, although it is not always clear if this number is for animals or animal equivalents such as SFU. For example, a 2003 World Bank report suggested 100 as a viable herd size. However, a 2009 World Bank report suggested 200 but did not specify the units, while FAO, UNICEF and UNDP (2007) suggested 100. These differences could reflect a change in average herd species composition with a switch from cattle and horses to small ruminants and, in particular, goats. Smaller herders, often more distant from administrative district centres (*sums*), tend to have less access to support services such as veterinarians. They are less well off and more likely to suffer food insecurity.

At privatization, very few herds had more

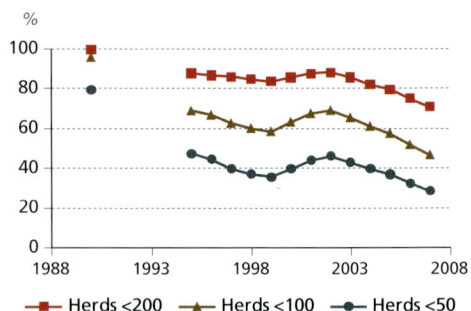

13 MONGOLIAN LIVESTOCK HERDS BELOW THREE CRITICAL SIZES

Source: Annual yearbooks (1998-2008) of the National Statistical Office of Mongolia with some calculations by authors.

than 100 animals and only 20 percent had more than 50. By 2007, 45 percent of flocks were less than 100, around 30 percent were less than 50 and only 30 percent were above 200. There was variation between 1995 and 2007, with herd sizes falling during the 1999–2002 *dzud*-drought combination and then recovering.

Nevertheless, even after five years of relatively good conditions between 2002 and 2007, by some measures almost half of the herds were too small to withstand the next period of climatic stress reliably, as shown in Figure 13.

It appears that herders with smaller numbers of animals are being gradually forced out of herding, a trend that continues even during relatively good years. Among those who remain as herding households, many are acutely vulnerable to poor climatic conditions and are likely to face periodic food insecurity, while the former herders are now counted among the rising number of urban households.

Mongolia is increasingly urban. Recent studies on food security have focused on urban households in Ulaanbaatar and provincial (*aimag*) centres where almost all urban dwellers are found. As much of the urbanization in recent years has been due to a push away from the rural areas, the same problems of lack of infrastructure, access to resources and food insecurity are seen in Mongolia as in other countries. A recent report (FAO,

UNICEF and UNDP, 2007) referred to the relative food security of herders and the common practice of family support for poorer households in the smallest urban centres, and contrasted this with the relatively greater food insecurity in the *aimag* centres and Ulaanbaatar where underemployment and unemployment are common, the heating cost for a *ger* in winter is high, and there is a lower intake of animal products and a greater reliance on cereals and potatoes for energy intake. A Mercy Corps study (Hillbruner and Murphy, 2008) found that around one-quarter of households in the *aimag* centres were moderately or severely food insecure, with a further 10 percent somewhat food insecure.

THE FUTURE OF FOOD SECURITY IN MONGOLIA AND THE CONTRIBUTION OF LIVESTOCK

While overall food supply in Mongolia is adequate, there are clear issues of distribution and access (due to poverty) and stability (due to climate, seasonal employment and urbanization). At a minimum, herders with less than 50 animals are at substantial risk of food insecurity while those with less than 100 have some risk. These two groups were respectively around 10 percent and 16 percent of all households in 2007. At the same time, 60 percent of households are urban and of those, 25 percent are food insecure. Thus, by combining the two herder groups with the urban, a total of around 25–30 percent of all households in Mongolia can be said to be food insecure.

Nomadic livestock keeping is a highly sophisticated and evolved system for making a living from a difficult environment. If traditional Mongolian herders still use systems and artefacts that are recognizable from historical tales, it is because they are very well adapted to the nature of the land and climate. Changing or "improving" such production systems is difficult. Bringing in external inputs can help, but this must be sustainable and not lead to a degradation of the environment on which the livestock system depends. Every animal that grazes needs a certain amount of feed biomass

to grow, reproduce and, importantly, to build up fat reserves for the winter. Although fodder is conserved for winter feeding, it has always been more common to conserve the excess summer grass growth as fat reserves on the animal, rather than as standing or cut hay. Biomass production is limited by soil fertility, growing season and rainfall. In Mongolia, the first two of these are limited and the third uncertain. The key factors are water supply in the summer and winter camp sites that have access to water and provide shelter but also have enough exposure for snow to blow away. A DANIDA report from 1992, quoted by Honhold (1995), estimated the total biomass production from Mongolian rangeland to be around 380 kg per hectare, sufficient to support 62.5 million SFU, assuming a 50 percent utilization by livestock. However, such a figure does not address annual variations, which are likely to be significant. Given that irrigated and artificially fertilized cropping land, probably some of the better land, produces around 800–1 000 kg of wheat per hectare, the figure of 380 kg of forage from un-irrigated land fertilized only by animal dung, would seem to be optimistic.

It is unlikely that extensive livestock systems can be adapted to produce enough protein to feed the country's growing human population, and there are limited prospects for establishing intensive systems. Livestock are still important in food supply but increasingly less so and, at the same time, livestock production may be at the peak levels possible with the resources available. Much of the country is remote, but little of it is wilderness untouched by human use. Some of the increase in livestock production has been achieved at the expense of once extensive herds of wild antelope. Livestock still contribute to the stability of income, with movements of people to and from herding due to different crises and shocks, but as herding household numbers fall in relation to urban households, this buffering effect has limits. Herding families are now only 28 percent of households.

Livestock dependant as it is, Mongolia relies on

imported food for its food security, either as grain or the inputs required for growing grain in a hostile environment. There is increasing reliance on imported foodstuffs or the inputs needed to produce them locally (although there are also limits to domestic crop production). The move to goats and the income they provide from cashmere has increased incomes and hence access to imported foods. However, Since most is exported mainly through informal channels, income depends on (volatile) world market. The export potential for other livestock products is probably limited because local demand is high. Export would create the need to import other products as substitutes. The animal health situation limits the export of live animals and most livestock food products. The recent opening up of large mining ventures, often with a significant government stake, may provide a source of income with which to import food, since profits are expected to contribute to a sovereign wealth fund for the country that will be used to support the population.

PROSPECTS FOR LIVESTOCK DEPENDENCE

Livestock-dependent societies, or those that are nearly so, play an important role in the contribution of the livestock sector to global food security. By supporting their own population and generating some surplus for export, they contribute to the world's supply of livestock protein as well as their own access to food.

However, the total production from these societies has probably reached the limit. Production per hectare is close to or at the maximum possible under the prevailing climatic and soil fertility conditions, given that many factors affecting production are beyond the control of livestock owners. The total area in the world available for extensive grazing is unlikely to expand because of competition for land from agriculture and human settlement, and therefore total production is likely to reach its limit sooner than in other systems. Existing levels of production should be protected to the extent possible because of their contribution to the food supply

and the protein balance, but the percentage that these societies contribute to global food supply can be expected to fall.

There may be shifts in location in the future brought about by climate change, which Black *et al.* (2008) describe as "one of the defining challenges of the 21st Century," one that is expected to change the shape of livestock production in Australia and perhaps in other countries where extensive grazing is widely practised. Decreased and more variable rainfall may require changes in management to cope with additional instability, while also creating new challenges for animal health systems.

Investment in market access will be important since this offers the potential for livestock owners to gain greater value from what they produce and to manage risk by managing stocking levels. The highest income comes from export markets for live animals, meat and fibres, but they are also volatile and particularly difficult for the poorest to access. Here the government has a role to play at national and international levels. For example, in Mongolia, if further development of the cashmere market were possible, it could increase the potential of the livestock sector to support food access. In the Horn of Africa, Aklilu and Catley (2009) suggest that regional policy frameworks within the IGAD and Common Market for Eastern and Southern Africa (COMESA) regional groupings could be supportive of livestock herders, including the poorest, by exploring a range of options for trade.

Over time it is likely livestock dependant societies will become less dependent on livestock, with their animals supporting and supported by other activities. There is a gradual trend for people to move into towns and away from pastoral agriculture. For those who choose to remain in rural areas, tourism, recreation and payment for environmental services such as wildlife conservation and carbon sequestration into grassland (as explained in more detail later) all offer ways to earn income that are complementary to livestock keeping.

©FAO/Ami Vitale

Small-scale mixed farmers

Almost every country in the world has communities centred on mixed farms with a diverse portfolio of activities that includes crops, livestock, other farm enterprises and non-agricultural work. A practical definition of a mixed farm is one where more than 10 percent of the dry matter fed to livestock comes from crop by-products and stubble or more than 10 percent of the value of farm production comes from non-livestock activities (Seré and Steinfeld, 1996). They are highly variable in terms of their size and location, the wealth of their owners, the way animals are managed and the part livestock play in food security. Mixed farms are estimated to produce the majority of the global meat and milk supply (48 percent of beef production, 53 percent of milk production and 33 percent of mutton from rain-fed mixed systems according to Steinfeld *et al.,* 2006).

Given the heterogeneity of the group, it is meaningless to generalize. Thus, this report focuses on the subset of mixed farmers for whom food security is least assured – those who live in developing and transitional economies and have small farms. In these countries, it is common to find communities where mixed farming, mostly small farms, predominates as a way of life.

Even among smallholder mixed farmers, there is still considerable variation in assets, income and social customs. However, a characteristic common to all of them is that livestock are managed as part of an integrated and tightly-woven system, in a way that fits the needs of the farm family, the available labour and the demands of other enterprises. Animals provide food, income, traction, manure, social capital, financial assets and a means of recycling crop wastes, all to varying degrees in varying situations. This is similar to their role previously described in livestock-dependent societies, but for mixed farmers, livestock are usually a much smaller farmers, livestock are usually a much smaller part of the portfolio, albeit an important one.

As this chapter will describe, livestock bring value, versatility and resilience to mixed farming households, which are more robust and food secure with animals than they would be without them. At the same time, there is an important unanswered question about the role of small-scale mixed farms in the food security of the future. These farms support the families who own them and provide extra food for local communities, but they offer limited prospects for supplying growing urban populations and limited opportunities for the economic advancement of farm households. They have the biological potential to produce a larger supply of food, and they produce food in ways that are positive for the edible protein balance, but there is little economic incentive for them to expand their production.

CONTRIBUTION OF LIVESTOCK TO FOOD SECURITY

Many farmers in rural areas survive by managing a mix of different crops and livestock activities, creating synergy when crop residues are used to feed animals and the manure from the animals is used to fertilize the crops. The different enterprises may be concentrated into the same small space or on separate farm plots. Other forms of mixed farming include grazing under fruit trees to keep the grass short or using manure from pigs to "feed" a fish pond. The prevalence of mixed farming varies by country and region. Figure 14, using figures from the RIGA dataset, shows an analysis from 14 countries where the proportion of rural households that practice both cropping and livestock ranges from 24 to 87 percent. Ly *et al.* (2010) reported that in 2004, 83 percent of the cattle in West Africa and 75 percent of the small ruminants were kept in mixed crop-livestock systems, with traction being an important reason for keeping cattle. Chacko *et al.* (2010) reported that from 2004 data, 83 percent of agricultural land in India was occupied by mixed farming systems.

Livestock contribute to food availability, access and stability. In some cases, direct provision of food is their primary contribution while, in others, the main motivation for keeping them is income. A rural household in India or Tanzania with one or two dairy animals will use the majority of their milk for home consumption (Garcia *et al.*, 2003; Knips, 2006). In Viet Nam, poor

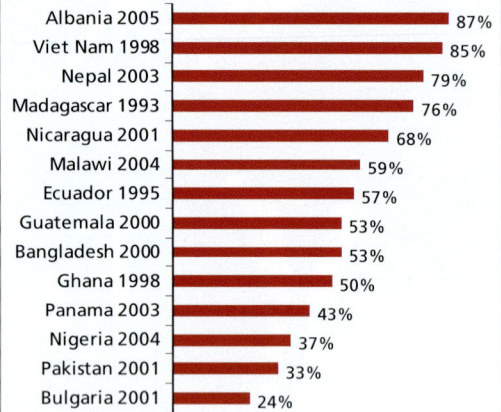

14 RURAL HOUSEHOLDS IN SELECTED COUNTRIES ENGAGED IN MIXED FARMING

Albania 2005	87%
Viet Nam 1998	85%
Nepal 2003	79%
Madagascar 1993	76%
Nicaragua 2001	68%
Malawi 2004	59%
Ecuador 1995	57%
Guatemala 2000	53%
Bangladesh 2000	53%
Ghana 1998	50%
Panama 2003	43%
Nigeria 2004	37%
Pakistan 2001	33%
Bulgaria 2001	24%

Source: RIGA dataset.

households that own small numbers of poultry as scavenging flocks use them mostly for home consumption (Maltsoglou and Rapsomanikis, 2005), while peri-urban poultry keepers are more likely than those in remote rural areas to keep flocks of sufficient size to have birds and eggs for sale (Hancock, 2006). In the countries shown in Figure 14, livestock's contribution to the income of mixed farming households ranges from a very small percentage to over 30 percent, with no consistent pattern according to the wealth of the family. Other studies show contributions of up to 50 percent at any given time.

The asset value of livestock is important to household resilience and food stability because it provides collateral to expand or diversify farming operations and gives households a capital item that can be sold in times of great need. Access to both formal and informal credit can be facilitated through ownership of livestock. A recent report found that in the countries represented in the RIGA dataset, livestock farmers were more likely to get credit from formal sources than non-livestock-keeping households within the same income bracket (Pica-Ciamarra, *et al.*, in preparation). The authors found this surprising, because in developing countries, unlike the more developed financial markets, "moveable assets" such as livestock are rarely used as collateral for formal loans. They concluded that livestock might act as a "buffer stock", allowing farmers to allocate part of their resources to relatively risky but high-return activities which financial institutions are willing to finance. For example, in Kenya, Imai (2003) found that that having a higher value of livestock assets enables households to invest more into high risk activities such as coffee and tea production. Another use of assets is to sell them for income smoothing – at times when other enterprises are not providing income or when the household is in crisis. Small animals provide more flexibility than large ones in these cases, since they do not require their owners to liquidate such a large proportion of their capital.

Gender affects the contribution of livestock in mixed farming households. In most developing and emerging economies, ownership of livestock is less common in female-headed than male-headed households. In the 14 countries shown in Table 10, only three have a higher percentage of livestock ownership in households headed by women. In spite of this trend, many examples of women contributing to food security can be found in mixed farming communities. Medium-sized duck breeding enterprises near Viet Nam's capital city Hanoi are equally likely to be owned and managed by women or men, and represent an important asset and income source for the household. Women are central to many of the dairy projects in India and East Africa, including Operation Flood (Arpi, 2006) and the Food and Agriculture Research Management-Africa's (FARM-Africa's) dairy goat project in Ethiopia (FARM-Africa, 2007) which has trained women as well as men animal health workers in recognition of a predominantly female clientele.

Two features distinguish livestock's role in small-scale mixed farming households from the contribution they make in other situations: the synergy between livestock and other enterprises, and the diversity and flexibility that livestock bring to the household's activities.

Synergy. Synergy with crops exists through the exchange of draft power, manure, pest control and crop residues. For example, herded ducks in the Mekong Delta and China travel from field to field eating snails, insects and discarded grain, thus providing pest control for rice crops (Yu *et al.*, 2008). As previously described, the use of draft power is widespread throughout the world although it is diminishing in most areas except Africa, where it appears to be increasing. In some cases, larger landholders own animals that smaller landowners share or contract for their use. Animal power allows the cropping area to be extended beyond what would be possible with hand cultivation, and allows land to be ploughed when it is dry in preparation for planting immediately after the first rains. Manure is most likely to be used for crops where animals and crops are in close proximity, although as explained previously, there are competing demands for manure and it can be in short supply.

Synergy with other livelihood enterprises is most evident with scavenging livestock. Income from these animals is low, but they often provide "something for nothing" by eating crop residues, insects, scraps and rubbish found within the community and requiring very little labour, equipment or housing. Scavenging poultry can provide a 600 percent return on the tiny investment they require (Otte, 2006). Scavenging pigs in Asia and Africa live on household waste, acting as garbage disposal units, and are housed at night in a rough shelter or kept in or under the family dwelling. Goats in Nepal live by grazing and on forages cut from communal grazing areas and forests, costing little in money although they demand time from women and children (ADB, 2010).

Diversity and flexibility. The contribution of livestock to food security varies over time depending on family need which can be for daily nutrition, for dealing with a food crisis or for developing a more solid economic base in which food security is assured. Poultry are particularly flexible because they have dual usage (meat and

TABLE 10

PERCENTAGE OF MALE AND FEMALE HEADED RURAL HOUSEHOLDS OWNING LIVESTOCK IN SELECTED COUNTRIES

		EXPENDITURE QUINTILE				
	HOUSEHOLD HEAD	1	2	3	4	5
Ghana 1998	Female	68	67	63	53	48
	Male	39	37	29	38	27
Madagascar 1993	Female	63	72	73	54	62
	Male	77	85	84	80	78
Malawi 2004	Female	49	58	64	61	59
	Male	63	74	73	74	66
Nigeria 2004	Female	26	25	24	31	32
	Male	50	49	47	43	39
Bangladesh 2000	Female	31	40	43	47	55
	Male	31	34	40	44	52
Nepal 2003	Female	67	86	73	73	75
	Male	81	87	87	85	83
Pakistan 2001	Female	52	49	58	54	54
	Male	57	62	63	67	66
Viet Nam 1998	Female	81	88	82	84	82
	Male	95	95	93	89	82
Albania 2005	Female	87	74	71	85	58
	Male	89	88	93	96	89
Bulgaria 2001	Female	27	46	73	77	75
	Male	34	67	76	78	73
Ecuador 1995	Female	76	80	78	79	69
	Male	69	72	79	68	74
Guatemala 2000	Female	67	71	63	58	52
	Male	68	72	70	67	57
Nicaragua 2001	Female	88	27	71	50	89
	Male	78	83	58	89	67
Panama 2003	Female	83	45	55	52	46
	Male	76	73	72	64	52

Source: RIGA dataset.

eggs) and can quickly be scaled up or down according to need. They have the fortunate characteristic of taking up little space so they fit well into peri-urban mixed farms – it is possible to keep 2 000 birds in a back garden. Larger flocks tend to be kept primarily for income and can be profitable when their owners have access to a well organized market chain (Ahuja *et al.*, 2008). In Southeast Asia, countries such as Indonesia, Viet Nam and Thailand have had a steadily growing demand for poultry. The gap in supply was first filled by small-scale entrepreneurs who moved quickly to meet a market need, but many of these producers left the market just as quickly when competition or government policies to control HPAI made market access more difficult (ACI, 2006; NaRanong, 2007).

Small ruminants also have short reproductive cycles and are particularly valuable where families have access to common grazing land or land where forage can be gathered and brought back to the animals. Small-scale commercial pig production fits well into mixed farms because it takes limited space and there can be some exchange of inputs between livestock and crops. In Viet Nam, crossbreds with indigenous pigs are not cost-effective to produce in large intensive units but are highly productive when fattened in small-scale units that hold 20–30 animals. Herds can be scaled up or down in a matter of weeks to meet demand cycles. They are such a delicacy that there is a thriving export of frozen piglet carcases from Viet Nam to Hong Kong (McLeod *et al.*, 2002).

CONSTRAINTS TO EXPANSION

The strengths of mixed farming also can be its weaknesses. The low-input, low-output systems that provide the family with "something for nothing" are efficient and effective in using waste, but poor producers of income or food. The intensively reared zero-grazed dairy cattle, dairy goats and small-scale commercial poultry and pigs common on peri-urban farms produce a higher output, but small land holdings and the need to diversify enterprises to spread risk mean that they tend to be small in scale and unable to benefit from certain kinds of new technologies. Indigenous livestock thrive under the conditions of mixed farms, which are often the best way of supplying niche markets. However, when small-scale farmers try to rear larger and faster growing crossbred and exotic animals, they cannot compete cost-wise with large and specialized commercial farms in the commodity markets to which these animals are suited.

Biosecurity measures. Keeping a mixture of livestock together within a small space makes it difficult to fully implement biosecurity measures. These are the physical and management barriers established to keep disease from entering or leaving herds and flocks. Under the best conditions, they require animals to be segregated by species and type and kept within fences or houses, call for keeping housing units a set distance apart, and restrict entry of people to the places where animals are kept. Lack of biosecurity creates a greater chance that animals will be exposed to disease. Lack of biosecurity measures also can prevent small-scale farmers from accessing lucrative urban markets that demand "certified safe" products.

Disease outbreak and control. If disease outbreaks do occur and control measures are implemented by the government to prevent disease from spreading, many farmers may suffer losses from culling (compulsory slaughter) of animals in and around the area of an outbreak, with small farmers more likely than large ones to have their animals culled without compensation (World Bank *et al.*, 2006). Imposition of quarantine measures also creates losses for small-scale livestock keepers, although traders may benefit from the prices that they can charge when the quarantine is lifted and animals flood the market (McLeod *et al.*, 2006). Expectation of losses from disease and disease control are built into the way the system is managed, often by keeping local animals that are better adapted to local conditions but produce less.

Resource scarcity. Small mixed farm households tend to be resource-constrained. Land is often in short supply, and many farming families are caught in a "poverty trap" when the small size of their landholdings restricts their access to credit and prospects for expansion. Family labour is often limited and hiring fulltime labour requires a certain scale of enterprise. Labour constraints are particularly noticeable when estimates of production are disaggregated by the sex of the head of household. In the countries represented in the RIGA dataset, the fact that female-headed households are less likely to be engaged in livestock farming than male-headed households can be interpreted as partly a labour constraint, since families with more working women own larger herds (Pica-Ciamarra *et al.*, in preparation).

Feed supply. In many countries, good quality feed is in short supply, which is a major constraint to expanding livestock production. In the State of Orissa, India, for example, even though buffalo-based dairy farming offers the lowest net milk production costs, it hardly exists because of the scarcity of feed sources (Garcia *et al.*, 2004b). Where possible, poor farmers will use agricultural by-products instead of commercial feed (Upton, 2004), but these can be limited. In India, even though poultry is an important source of protein for home consumption, backyard poultry producers cannot increase production because of the limited availability of scavenge-based feed sources (Pica-Ciamarra and Otte, 2009). Cereals in Africa and Asia are often contaminated with aflatoxin (Hell *et al.*, 2008), meaning that commercial companies prefer to import cereal for their compound feeds.

Improvement costs. While commercializing or scaling-up livestock production can be seen by outsiders as an attractive option to improve income for mixed farmers, those proposing it often fail to appreciate the extra effort and expenditure involved. Transforming a scavenging system into one where animals are entirely or mostly confined can greatly increase their output but, at the same time, will greatly increase the cost of housing, feed and animal health care plus the time that is spent caring for the animals. Acquiring a high-value, high-producing animal such as a crossbred dairy goat or dairy cow requires a large up-front investment in a shed and incurs recurrent expenses of feed and health care, as well as the need to be connected to a reliable market to sell the extra produce. For this reason, NGOs such as Heifer International and FARM-Africa which run small-scale dairy projects always require farmers to be very well trained and prepared before they receive an animal.

Small-scale mixed farming is found throughout the world, in both developed and developing countries.. As shown in the examples mentioned in this chapter, no one country is representative of all. However, the following case study of Nepal provides a good illustration of several of the issues raised in this chapter. It looks at the contribution small-scale mixed farming households make to the economy of Nepal, the constraints they face and the part livestock play in the food security of these households.

CASE STUDY
MIXED FARMING IN NEPAL

Of Nepal's population of 29.1 million, around 80 percent live in rural areas, 79 percent of whom practice mixed farming. As with many other countries, Nepal is urbanizing. In 1985, just 7 percent of the population lived in urban areas, compared to 20 percent in 2001. The rate of out-migration to other countries is also increasing (FAO, 2009a), mostly to India, the Near East, Malaysia and the USA. Nevertheless, mixed farming is still a very important contributor to livelihoods, with agriculture providing more than one-third of GDP (39.1 percent in 2001) (Maltsoglou and Taniguchi, 2004).

Mixed farming is carried out in conditions of poverty and intermittent social instability. Nepal is one of the poorest countries in the world, ranked 99 out of 135 countries in the Human Poverty Index (UNDP, 2009), and has become a food-deficit country. In 2006, 4.2 million people, representing 16 percent of its total population, were undernourished (FAO, 2009a). A recent government report found that 3.35 million people and 40 percent of the population in the mountain and hill districts were facing a severe food crisis (Kharel, undated). The nutritional status of mothers and children under five is extremely poor. There is no or very restricted basic infrastructure in rural and peri-urban communities, and social services such as medical care, clean drinking water and adequate sanitation are very limited. Although agriculture is very important, the performance of the sector has been inadequate to meet increasing food demand, and low agricultural productivity is a major cause of food insecurity.

LIVESTOCK IN THE SYSTEM

The country is divided into three main geographical and ecological regions, the mountain region, the mid-hill region including the Kathmandu valley, and the Terai (lowland) region.

TABLE 11

NUMBER OF LANDOWNERS AND LANDLESS HOUSEHOLDS BY GEOGRAPHICAL REGION OF NEPAL

	EASTERN	CENTRAL	WESTERN	MIDWEST	FARWEST	TOTAL
Hhs owning land	462	604	424	292	308	2 090
Hhs not owning land	184	189	109	35	3	520
Total	646	793	533	327	311	2 610

Source: RIGA dataset for 2003–04 (survey of 2610 households).

The mountainous and hilly regions are quite isolated since road access can sometimes be very difficult. The average time to travel from a mountain farm to a health post or a primary school has been estimated between 1.8 and 2.2 hours. The Terai as well as Kathmandu and the other urban areas have better road connections, and the Terai is generally very accessible.

Mixed farms are found in all regions but livestock play a more central role in the mountain region where, because of harsh cold climatic conditions and infertile land, it is hard to grow crops. Animals are kept in low input, extensive systems (Parthasarathy and Birthal, 2008), and people are more dependent on livestock husbandry than in other regions. Households use livestock mostly for home consumption, especially in the mountains and rural hills due to their remoteness, but livestock are also an important source of the little cash the households in these areas earn. In the Terai and hill regions, about two-thirds of livestock keepers are smallholders (Gurung *et al.*, 2005) and most of them are mixed farmers.

A high percentage of households in rural areas own land (80 percent), but most landholdings are very small (Figure 15), with plots becoming further fragmented as they are divided up for inheritance. No major differences in land size are found in the different areas of the country or among households of different expenditure quintiles. In the far west, almost everyone owns land, with ownership decreasing progressively from west to east, reaching the lower limit,

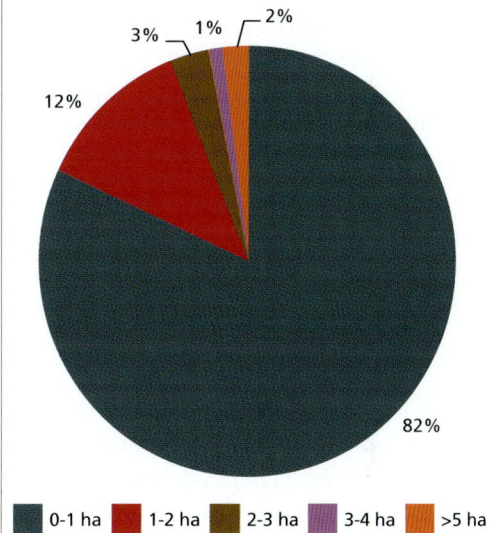

15 LAND SIZE AMONG LANDOWNER HOUSEHOLDS OF NEPAL

0-1 ha ■ 1-2 ha ■ 2-3 ha ■ 3-4 ha ■ >5 ha

Source: RIGA dataset for Nepal, 2003–04.

72 percent, in the east (Table 11).

Wealth does not influence whether rural households own livestock (Table 12) but it does influence how many they own. Almost every household has livestock of some kind, but landowners are more likely to own more than one tropical livestock unit (TLU) – which is equivalent to 5 pigs or 2 cattle using the international measurement for South Asian livestock – than those who are landless (Table 13). Herd sizes are generally very small. Mixed farmers tend to own more of each species than other households

TABLE 12

PERCENT OF NEPALI RURAL HOUSEHOLDS OWNING LIVESTOCK, BY EXPENDITURE QUINTILES

RURAL HHS	EXPENDITURE QUINTILES				
	1	2	3	4	5
2 610	87%	90%	88%	87%	86%

Source: RIGA dataset for Nepal, 2003-4.

TABLE 13

PERCENT OF LANDOWNING AND LANDLESS HOUSEHOLDS BY TROPICAL LIVESTOCK UNIT (TLU) OWNED

TLU OWNED	LANDOWNER HOUSEHOLDS	LANDLESS HOUSEHOLDS	TOTAL HOUSEHOLDS
0	1%	1%	1%
0-1	14%	33%	18%
>1	85%	66%	81%
Total	100%	100%	100%

TLU estimated using international units for South Asian livestock.
Source: RIGA dataset for Nepal, 2003–04.

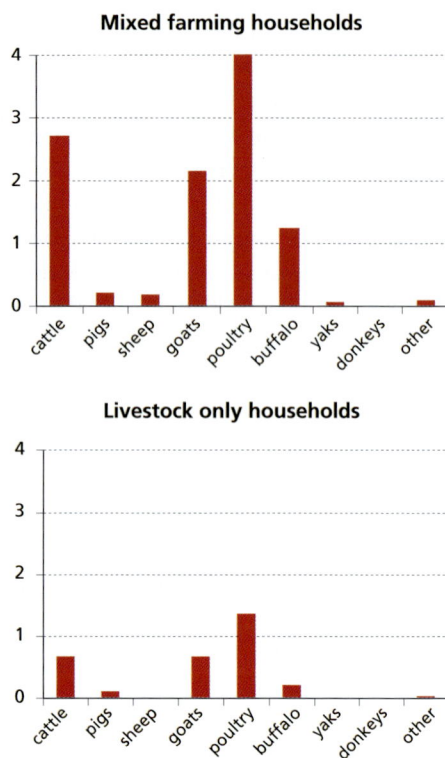

16 TYPE AND AVERAGE NUMBER OF LIVESTOCK OWNED BY MIXED FARMING AND LIVESTOCK-ONLY HOUSEHOLDS IN NEPAL

Source: RIGA dataset for Nepal, 2003–04.

(Figure 16) but the average number of TLUs owned is around two, regardless of the wealth of the household (Table 14). Female-headed households tend to own fewer animals, averaging 1.2–1.8 TLUs compared to 1.9–2.0 TLUs for male-headed households.

Subsistence cropping is predominant, with households growing crops mostly to consume at home rather than for sale. Only a few families with larger landholdings are able to produce in excess of their consumption requirements and profit from the sale of their products. Most small landowners have to seek alternative means of income and still face food shortages for several months of the year. Paddy rice is the most commonly produced staple, followed by coarse cereals and wheat. Pulses, oilseeds, vegetables and potatoes are also grown in smaller quantities. Livestock play the usual wide range of roles in a mixed farming society. They have an eco-

nomic role since they can be used for income and as insurance to hedge against risk. Mixed farming households in all wealth categories have a higher average income than non-mixed farming households (Table 15). Regardless of wealth, these households earn roughly 60 percent of

TABLE 14

DISTRIBUTION OF TLU AND TYPOLOGY OF LIVESTOCK, BY EXPENDITURE QUINTILES

EXPENDITURE QUINTILE	TOTAL TLU	NUMBER OF LARGE RUMINANTS	NUMBER OF SMALL RUMINANTS	NUMBER OF POULTRY	NUMBER OF PIGS
Poorest	1.94	3.39	2.07	2.76	0.23
2	1.96	3.52	2.18	3.54	0.22
3	1.80	2.99	1.83	3.27	0.16
4	2.01	3.40	2.06	3.22	0.14
Richest	1.96	3.09	1.97	4.16	0.21

TLU estimated using international units for South Asian livestock.
Source: RIGA dataset for Nepal 2003–04.

TABLE 15

INCOME TOTAL FROM LIVESTOCK AND CROPPING (IN NEPALESE RUPEES)

EXPENDITURE QUINTILE	TOTAL INCOME FOR NON-MIXED FARMING HOUSEHOLDS	TOTAL INCOME FOR MIXED FARMING HOUSEHOLDS
Poorest households	16 805	22 474
2	24 662	26 982
3	19 617	27 687
4	25 210	31 654
Richest households	35 721	33 621

Income = cash and production for home consumption.
Source: RIGA dataset for Nepal 2003–04.

their income from agricultural activities, of which livestock contribute close to 40 percent. This is similar to other Asian countries in the RIGA dataset, a level of around 30 percent being usual, while countries in all of the other regions have lower percentages.

Households without livestock are much more reliant on off-farm income and wage labour. The social and cultural role of livestock is also important, especially during ceremonies. Goats and chicken are kept to provide for guests and for religious purposes, as some ethnic communities believe it is necessary to sacrifice one goat and chicken every year (Gurung *et al.*, 2005). Livestock also provide status in the community and create employment opportunities within and beyond a given household.

SOCIAL INFLUENCES

Nepal is a pluralistic society with about 60 recorded caste and ethnic groups, and 70 languages and dialects (Gurung *et al.*, 2005). Ethnicity is an important phenomenon and, along with caste, is the most important focus around which individuals, households and communities aggregate and, as such, it influences livelihood options. Some 37 percent of the population is made up of indigenous "ethnic groups" outside of the caste system and 13 percent are in untouchable caste groups. Many of these groups have been historically disadvantaged and continue to lag behind in income and asset levels, educational achievements and human development indicators. Religious customs affect livestock ownership and the role of livestock in food security. For example, strict Brahmins do not eat meat, and Hindu castes do not rear pigs and consider them unclean.

The other social factor that affects livestock keeping is gender. Most mixed farming households (87 percent) are headed by men and, as in many other countries, male-headed households have easier access to land and to both formal and informal credit, which are important for households to buy livestock. Women and men traditionally have different responsibilities, knowledge and decision-making roles in livestock management as well as other intra- and inter-household activities. Women's major responsibilities lie in caring for poultry, collecting grasses and fodders, and feeding, cleaning and milking the animals. Women and children generally are responsible for small ruminants and poultry as well as pregnant and sick animals kept at the home. Men's specific roles include providing veterinary services, livestock investment and spending cash for various household economic and community activities. They generally care for and control more lucrative animals with higher commercial value, such as cattle and buffaloes. They are responsible for selling and often are the sole decision-makers on how to use the resulting income. Men and women share activities such as cropping, grass and fodder production, traditional livestock breeding and species selection (Gurung et al., 2005).

The picture that emerges is that livestock keeping is very much a joint activity. Men and women from all socio-economic groups and regions take part in caring for animals and selling their products. However, even though women have higher decision-making power with regard to small animals, their decisions over large animals, sale of products, investment and animal health care are very limited. Women often depend on men (husbands or male relatives) for access to land and other inputs needed for more productive agriculture. However, the responsibility of women in many livestock management activities has been growing in the past few years (Gurung et al., 2005). With increased migration of men to cities or to other countries, agriculture has "feminized" in rural Nepal, meaning that women must often take on new responsibilities – often with limited knowledge, technology and time. This suggests that the need for men to seek income outside of their home communities may be another factor limiting the productivity of mixed farms.

THE FUTURE FOR MIXED FARMING AND THE CONTRIBUTION OF LIVESTOCK

Within the options available in a very poor country, mixed farming activities appear to be a winning strategy for rural Nepalese. They provide higher income than can be obtained by families relying on wage labour and off-farm employment, and they offer a measure of stability and control. Mixed farming is possible for many rural families, in part because of the very high rate of land ownership. Livestock make an important contribution to income and social functions. There is also a growing demand for food, including livestock products, in the urban population.

However, the opportunities for mixed farming to provide a pathway out of poverty, or to expand its production, are very few. Farm sizes are extremely small, and much of the country is on hilly terrain that requires huge efforts to farm, meaning that even if families decided to combine efforts, they would find it hard to upscale. In the Terai, there is more chance to combine landholdings and upscale because the terrain is flat, and the climate and water supply are more favourable. However, increases in productivity through upscaling could only come about through change in land tenure, either through a reduced number of people owning land or through cooperative arrangements in landholding, neither of which is likely to be socially acceptable at present.

Most families in Nepal have very limited access to the level of funding that would allow them to innovate. The migration of men away from rural areas also reduces the labour force. At best, mixed farming has the prospect of continuing to provide food security for the communities in which it is practised, with a small surplus being exported to towns. Livestock contributes by adding stability to the system and providing much of the cash that can sustain families when crops are not there. Some small possibilities exist to increase livestock productivity through slightly improved feeding, better veterinary care and more organized marketing. Technical options for accomplishing this have been explored within the national research systems and development projects but, only very small incremental changes to improve productivity, based mainly on better rural service provision have been recorded.

PROSPECTS FOR SMALL-SCALE MIXED FARMERS

Small-scale mixed farms remain enormously important because of the large number of rural households they support. They also make a useful contribution to the food supply of urban populations in developing countries and use and recycle resources effectively.

At the same time, they have limited prospects to expand their production or increase productivity. The Nepal case study provides a sharp reminder of the reasons for those limits. Lack of opportunity or capital to increase farm sizes, limited assets and therefore limited access to credit, lack of investment capital, limited land availability, reduced access to common land, higher unit costs than those of large producers, and restricted opportunities to market produce through physical distance or barriers imposed by quality and safety requirements are all factors that prevent many small-scale farmers in many places from expanding or intensifying their production. These factors constrain the stability of their own food security and the extent to which they can contribute to national food security.

All small-scale farmers do not face all of the same constraints. For example, in peri-urban India and Kenya they have excellent connections to milk markets, as previously described. In Kenya and Uganda, they face severe constraints on land size but have benefited from improved fodder species, and in some cases, access to animal health service through projects or co-operative arrangements. However, most small farmers face limits to intensification. Even in India, where small-scale livestock producers are supported by state investment, the average number of poultry kept by farmers with 0.5 to 2 ha has grown much more slowly than that kept by farmers with more than 4 ha, and the number of cattle has declined slightly on the smaller farms and increased slightly on large ones (Chacko *et al.*, 2010). Many successful small-scale peri-ur-

ban farmers in Africa have other jobs – including civil service jobs – and do not rely on mixed farming for their food security.

There continues to be large growth in demand for pig and poultry products, but small-scale producers of these products face strong competition from large-scale producers with intensive farms that may specialize only in livestock. One estimate suggests that large intensive farms produce 67 percent of world poultry meat, 50 percent of eggs and 42 percent of pork (Blackmore and Keeley, 2009) and they do so with a cost efficiency that small farms find hard to match. Pig and poultry production is scaling up and market chains are becoming more integrated in emerging economies such as Brazil, Costa Rica (Ibrahim *et al.*, 2010) and China (Ke, 2010). In 1996, fewer than 20 percent of pigs in China were produced on large farms, but by 2006, the figure was 64 percent.

Expansion in demand will increasingly come from cities. Peri-urban small-scale farmers tend to be very successful at supplying urban populations in the early stages of demand growth but less so as food safety and land use regulations become stricter, a topic explored in later chapters. Countless reports propose ways to connect small-scale farmers to markets (LPP, 2010), but if they are to continue to access the market chains that supply large towns and cities, they need to be credible competitors. For some, it is possible to become contract farmers to larger operations (Gura, 2008; Delgado *et al.*, 2008). For others, innovative approaches may offer the chance to access niche markets (Ifft *et al.*, 2009), perhaps through cooperative arrangements. For the remainder, in fast-growing developing countries, "it is hard to see a bright future" (Delgado *et al.*, 2008).

What we may see is increasing heterogeneity among small-scale mixed farms, with some, particularly in rural areas, remaining integrated operations with a mixture of crops, livestock and other enterprises and, within their livestock enterprises, a mixture of scavenging herds and flocks and small-scale intensive units. They will never earn a large income but will remain important contributors to food supply and access for the communities and local markets they supply. Livestock will continue to be essential to these systems although, in some places, they may be overtaken by aquaculture. For others, the best short term option may be contract farming. This implies a shift towards specialization with a smaller number of enterprises, each representing a higher proportion of assets and income. The Chinese poultry sector, for example, still presents good opportunities for contract farmers, providing around 800 000 jobs (Blackmore & Kelley, 2009; Ke and Han, 2007). Contract farmers often earn more income than their independent counterparts, but the stability of their income and hence food access may become better or worse depending on the contract. During the 2005–06 HPAI outbreaks, it was reported (personal communication with various people in the sector) that some Thai contract poultry farmers were protected from losses because the companies owned their birds and re-supplied them as soon as the immediate crisis was over. On the other hand, during the 2007–08 economic crisis, some farmers supplying supermarket food chains lost contracts very suddenly when activities were downsized.

Over time, although it is hard to predict what the time scale will be, we can expect to see a reduction in the number of small mixed farms worldwide, faster in some places than others. As they decline in number, the communities on which they are based will change in character, becoming less dominated by the agricultural calendar and more by the demands of other employment. They may also become more stratified, with some farmers continuing to exist just above the poverty line, some leaving farming for other employment, some becoming financially successful through contract farming, and a few managing to upscale or succeed in niche markets.

City populations

By 2007, half of the world's population was living in urban areas (UNFPA, 2007), a considerable increase from 29 percent in 1940. The developed world (North America, Japan, Europe, Australia and New Zealand) is highly urbanized, with 75 percent of people living in towns and cities, while in countries defined by the UN as "least developed", the figure is 29 percent but climbing (UNFPA, 2009). This is an important development that affects food supply systems, since urban populations are to a large extent purely consumers of food, unlike those in rural areas who both produce and consume it.

Those responsible for planning and managing urban spaces aim to ensure that stable supplies of reasonably priced food are available for all, through food chains with high standards of hygiene and safety. FAO (2001) identifies areas of concern for urban food supply and distribution:

- food supply – must be sufficient in quantity and quality, produced in hygienic and environmentally sound conditions and brought to the town or city by an efficient transportation system;

- food distribution within the town or city – requires investment by the public and private sector as well as legislation and regulations; and

- health and the environment – includes protection of the air and water supply and the health of people.

Each of these elements tends to be planned and managed differently in industrial market-based economies, centrally planned systems and market-based developing countries.

Urbanization affects the demand for food because urban people are, on average, richer than rural people and have access to food from a variety of sources. People living within or in easy reach of urban areas eat diets that are different from and more diverse than rural dwellers' diets (Regmi and Dyck, undated). However, there is an enormous range of wealth within urban populations. About 300 million urban dwellers worldwide are classified as extremely poor (Ahmed *et al.*, 2007), with the poorest people in cities highly food insecure. Countries with growing urban populations as well as rising wealth face the strongest challenge, because they must deal with two different food security problems – a large proportion of the population that is undernourished but also a growing number of people who consume more than they need for health or have poorly balanced diets.

The location of livestock production and the shape of livestock market chains are increasingly driven by urbanization and particularly the growth of large cities. This chapter compares approaches and experiences in feeding cities in the USA, Asia, Africa and Latin America.

LIVESTOCK PRODUCTS IN THE URBAN DIET

Urbanization has been associated with a rising demand for livestock products throughout the livestock revolution. Urban people, on average, eat less starch staples and more meat, fruit and vegetables than rural people (ICASEPS, 2008; Hooper *et al.*, 2008; Regmi and Dyck, undated). For the most part, this is because large towns and cities offer more income-earning opportunities than rural areas, and urban people are on average richer. However, poor urban dwellers eat far less livestock source food than their richer counterparts.

For those who can afford it, livestock products are highly accessible in cities. Fast food establishments, restaurants and large supermarkets sell livestock protein conveniently packaged at a wide range of prices. The large numbers of urban poor, however, have low purchasing power and limited food options, and are often physically separated from sources of quality food (Associated Press, 2008).

QUALITY AND SAFETY

Livestock products can be a valuable part of a balanced diet for city dwellers with sufficient incomes. However, many of these consumers who place convenience and immediate satisfaction over nutritional value are faced with the temptation of easily available livestock foods prepared in large portions and cooked with fat and salt, an incentive to overeat. The over-consumption of red meat and fats associated with heart disease and other health problems, mentioned in con-

nection with livestock food in the diet, is very much a problem of urban populations.

Some middle class consumers are highly discerning about their food and, when provided with sufficient information to give them confidence in the product, will choose food that they perceive to be safer or in some other way of higher quality, even if it is slightly more expensive (Birol, Roy and Torero, 2010). This translates into a demand for livestock products certified as having one or more of the following qualities: they come from livestock that have been raised traditionally, kept under high welfare standards or are biosecure, or they are from a particular breed or region or processed in a particular way.

While they represent relatively small numbers, these consumers have raised the standards demanded of livestock producers in Europe and other parts of the developed world, and in pockets in emerging economies or urban markets of developing countries. Food safety carries a high premium for these consumers, because even if they do not search for it, they are quick to retreat from foods that have been associated with outbreaks of human disease. Supermarkets, an important source of city food, are risk averse and pass part of the cost of food safety to their suppliers through demands for high levels of biosecurity and hygiene.

The urban poor, however, eat less livestock protein than their richer counterparts and their choice is restricted by the high prices of many foodstuffs. Food safety is a concern for them if food is delivered through long market chains in which hygiene, refrigeration and toxin and residue levels are not regulated or monitored. In developing countries, the government resources dedicated to food safety tend to be devoted more to quality control of export products than regulation of domestic food chains (FAO, 2009b). Food safety is a concern for the poor in general, but those in large cities have less access than their rural counterparts to local markets where they can purchase a live chicken that they have inspected for its health status, or determine the provenance and age of meat or milk. As a result, they depend more on the protection conferred by food safety regulations.

EFFECTS OF FOOD PRICE RISES

Poor people are vulnerable to food price rises, as previously discussed, because they spend a large proportion of their household budget on food. Poor people living in large cities are particularly vulnerable because they have weak connections to agriculture (Cohen and Garrett, 2010). They cannot do what mixed farmers do and change the balance of what they sell and directly consume to suit the prevailing economic situation. As the next section discusses, there are mixed farmers living within city limits, but they are far fewer in number than in rural areas.

Urbanization is contributing to the growth of demand for livestock products, but it also may be a minor contributory factor to the rising price of food since urban households are likely to hoard food if they fear future price increases (Stage *et al.*, 2010).

During the economic crisis of 2007–08, world prices of staples rose enormously, by three times for maize and five times for rice. World prices had a much stronger impact on domestic prices in some countries than in others (Cohen and Garrett, 2010, citing several sources), but the poor in many large cities cut back on food consumption and adjusted the composition of their diets. For example in April 2008, it was reported that poor households in Dhaka, Bangladesh, had stopped eating meat, fish and eggs (Cohen and Garrett, 2010) while in Ethiopia, they cut out eggs and vegetables. When food and cooking fuel costs rise, street food consumption tends to increase (FAO, 1997), as street food vendors can buy in bulk, while poor households buy in small quantities.

SOURCES OF LIVESTOCK SOURCE FOOD FOR URBAN POPULATIONS

There are three sources of livestock products for urban areas: the animals kept (often illegally) within city limits, peri-urban farms at the fring-es of cities, and large commercial operations delivering their product through integrated market chains that may span many miles and cross international borders. This section begins by reviewing livestock keeping within cities, a topic which is somewhat neglected in the literature. It then looks at the limits from which livestock source food is drawn into cities, the way that different governments approach the supply of city populations, and other factors that affect the shape of livestock market chains.

LIVESTOCK IN CITIES AND ATTEMPTS TO KEEP THEM OUT

Livestock have always been part of urban landscapes, but as cities grow and become more organized, the authorities try to exclude animal farms and slaughter facilities from residential areas and city centres because of concerns about human health, noise, dirt, smells, vermin and contamination of water supplies. These problems stem from pressure on land, meaning that people and their animals are forced to live in close proximity. Urban sanitation infrastructure is already strained and the poorer inhabitants, those most likely to want to keep livestock, often lack water, drainage and rubbish disposal facilities. Therefore, far fewer animals are kept in urban than rural areas, particularly in developed countries.

The history of livestock in cities of the USA (Box 7) has interesting parallels to stories elsewhere. During the early twentieth century, zoning codes, by-laws, regulations governing market chains and industry practice pushed livestock out of residential areas and city centres. In Kenya, Nairobi experienced a similar promulgation of laws restricting animal agriculture within city limits with by-laws dating from colonial times. The Agriculture Act, the Land Control Act and the Physical Planning Act offer local authorities the legal power to decide whether or not to allow urban farming. Yet, the legislation is rife with contradictions, and farm animals are still commonly found within city limits (Foecken, 2006; Foecken and Mwangi, undated).

BOX 7

CITY LIVESTOCK IN THE USA

In the USA, early urban planners integrated animal agriculture facilities with cities. In 1870, New York's Central Park incorporated a dairy barn on the premises as a way of providing the urban poor with milk during a time when transport to rural dairy farms was limited.

Yet the turn of the twentieth century saw a push to exclude farm animals from cities for a variety of reasons. Dairy cows were banned due to the health risk they posed to people from the spread of bovine tuberculosis (Schlebecker, 1967). Farm animals were seen as noise and waste management problems for cities. The birth of "animal welfare" activism created a push to move animals out of cities where they were not properly cared for. Chickens were banned with the excuse of preventing rooster fighting and as part of noise ordinances and anti-nuisance laws.

Most of the early zoning codes in the USA imposed bans on all "farmyard animals" simply to prevent noise and smell. Exceptions were granted for horses, which were widely used for transportation until the 1920s. Laws regarding food animals were usually not a state or city-wide regulation, but tended to be locally based. Each housing development could have different standards in their zoning ordinances and deeds. The first animal restriction listed in a table of many of the subdivision restrictions, prepared by H. V. Hubbell (1925), was an 1889 Baltimore County, Maryland, statute for "no pigs, allow fowls, four horses, and two cows." Some historians have speculated that the early bans in these early planned communities had a covert reason: to keep out lower income groups that would need animals as supplemental income.

However, the movement of animals out of early American cities was not wholly a factor of early zoning codes and by-laws. As industries achieved aggregate economies of scale starting in the early 1900s with meat markets and poultry, they may have influenced decisions to prohibit potential customers from rearing, slaughtering or selling animal products for private profit. For example, new laws requiring commercial dairies to sell pure milk drove smaller cow stalls in towns out of business simply because of the cost of testing the milk and the lack of space to expand. Other policies, such as immunity to anti-trust laws in agriculture, favoured larger producers and economies of scale over local, small-scale animal agriculture. Immunity to anti-trust laws gives larger companies advantages in terms of favourable marketing and packaging deals for greater quantities of goods. These laws are now being challenged and this may, in time, affect where animal agriculture is located by removing some incentive for large, contract-based farming operations (*The Economist*, 2010).

Despite the century-long bans on urban animal agriculture, practices coupling cities with animal agriculture have persisted. Philadelphia employed a peri-urban swine feeding consortium that consumed up to 1 500 tonnes of residential organic waste a week, as late as the 1980s (Maykuth, 1998). Such practices are still commonplace in cities outside of the USA. Walmart, an international food supplier, now considers garbage feeding part of its sustainability best practices (Walmart, 2010).

Source: Brinkley, 2010.

After decades during which city dwellers reared poultry in Jakarta, Indonesia, the Jakarta Province authorities passed legislation in 2007 and 2008 that banned poultry keeping within city limits except for certain licensed birds not reared for food, and initiated moves to close holding yards and slaughter points in parts of the city (ICASEPS, 2008). The reasons cited were related to HPAI control, but complaints from residents about smells and dirt seem to have added impetus. Another move to ban urban livestock occurred in Cairo in 2009, when the small-scale operations that recycled garbage through pigs were closed down (*The Economist*, 2009). In both of these cases, the overall intention of improving environmental hygiene was positive, but there were negative impacts on the livelihoods of poor city dwellers.

Regulations other than zoning, as well as economic factors, have influenced urban livestock keeping. In the early twentieth century, the UK's banning of swill feeding to prevent the spread of pig diseases quite rapidly led to the cessation of small-scale pig keeping, much of which had been done in the back gardens and allotments of town dwellers. In Thailand, tax incentives encouraged livestock producers to move away from Bangkok (Costales *et al.*, 2006).

Notwithstanding attempts to keep them out, livestock still can be found within and at the periphery of many urban areas throughout Africa, Asia, Latin America and the Near East. Poor households keep small livestock such as poultry, guinea pigs and rabbits on rooftops and in courtyards for their own consumption. In places where they are not prevented, animals scavenge in the streets or, as in the case of Cairo's former pig keepers, are kept as garbage recycling units. Immigrants to urban areas bring their animals with them to satisfy their taste for traditional food from their homelands.

Several studies in the 1990s showed the prevalence of livestock in and immediately around African cities. An average of 17 percent of the inhabitants of six Kenyan towns were keeping livestock (Lee-Smith and Memon, 1994), and

the cattle population of Nairobi was estimated at 28 000, with most animals kept for manure and as savings accounts. However, the larger the town, the smaller the proportion of its population that engaged in agriculture of any kind. In and close to Kampala, Uganda, around 25 to 30 percent of people kept livestock (Maxwell, 1994), a tradition that appears to have persisted (Lee-Smith, 2010, citing studies from 2003). In Ghana, 25 percent of small ruminants were kept by people in and around urban areas, and in Mali, there were small communal dairies in the capital city, Bamako, in 1993 (Debrah, 1993).

There is a thriving small-scale poultry industry in and at the fringes of cities in Asia and Africa. In Cairo, small commercial units of a few hundred birds (FAO, 2009c) kept in narrow passageways play an important part in feeding the city's inhabitants. In Indonesia, Jakarta had an estimated 194 200 head of poultry in 2003 and 175 000 in 2007 (Directorate General of Livestock Services, 2007, cited by ICASEPS, 2008) although the number and flock size reduced after the 2007-08 government bans on poultry keeping.

City livestock are more important to the food supply than is sometimes acknowledged. However they only represent a small part of the whole. The next section discusses the diversity of livestock market chains that supply cities and the way that policies have contributed to shaping them.

FOODSHEDS, CITY LIMITS AND LIVESTOCK MARKET CHAINS

Two of the important factors that define a market chain are its physical length and its concentration, meaning the number and scale of units at each link of the chain. Urban planners talk about the "foodshed" – the area around a city that can conveniently provide food for its inhabitants. In the USA, the foodsheds of Philadelphia and San Francisco are defined as a radius of 100 miles from the city centre. Recent studies indicate a highly varied food system, with Philadelphia sourcing nearly 50 percent of its food from the foodshed and exporting 36 percent of production from the area, while San Francisco's total food demand accounts for only 5 percent of production within the 100 mile radius, with most of the production from its foodshed exported (Thompson *et al.*, 2008). Both the San Francisco and Philadelphia studies indicate that, despite an abundance of peri-urban farming, the cities still draw significantly from the national and international food systems.

These American cities indicate a disconnect between markets and local production, similar to the situation in Belo-Horizonte, capital of Brazil's Minas Gerais State, where the municipal government has invested in partnerships with the private sector, established marketing regulations and developed programmes to support local peri-urban production as well as incentives for consumption of local foods. In Mexico City, mobile markets have been set up that move around the city on specific days and often sell local products.

The Chinese government has taken a very different approach to that of the USA. The foodsheds for large Chinese cities are defined by their city limits. They aim and partly succeed (Girardet, 1999) in being as self sufficient as possible within these limits. This, in turn, has affected their zoning regulations and definitions of city limits. The official boundaries for Chinese mega-cities are larger than city limit lines in much of the rest of the world. The preoccupation with self-sufficiency is partly attributable to changes in city boundaries under the Great Leap Forward policies of the late 1950s that emphasized making the major Chinese cities self-reliant in food.[3]

Beijing increased in land area from 4 822 km² in 1956 to 16 808 km² in 1958, thereby incorporating much peri-urban agriculture under the city's direct control. Within the Beijing city limits, "urban agriculture" supplies 70 percent of

non-staple food to city inhabitants, mainly consisting of vegetables and milk (Jianming, 2003). Shanghai has taken a similar approach (Box 8) by defining an area for its "city limits" that is only 13 percent urban. It produces biogas energy as well as food within that area, thereby making a contribution to dealing with pollution from manure, a huge problem when livestock are concentrated close to large cities.

Within large African cities, while the city limits may not be defined as widely as those in China, Lee Smith *et al.* (2010) talk of an "agriculture gradient", with a relatively small number of city farmers near the centre and a progressively larger number towards the periphery and in the surrounding peri-urban area. Surveys do not always define clearly where they assume the city limits to be, which makes it difficult to compare statistics. In some cases, there is a deliberate policy to support urban farmers, as in Kampala where 26 percent of households within urban zones and 56 percent in the peri-urban zones were practicing some kind of agriculture in 2003. Summarizing findings from several papers, Lee Smith *et al.* (2010) suggest that livestock keeping within city limits is beneficial to food security in the city, but may be less beneficial to the poorest households than to richer ones that have better access to urban land.

Notwithstanding the uneasy relationship between livestock and cities, quite a large proportion of livestock product comes from within or close to city limits. FAO estimated that 34 percent of total meat production and nearly 70 percent of egg production worldwide came from peri-urban farms in the late-1990s (FAO, 1999). In the USA in the early 1990s, counties defined as urban influenced, meaning those within or adjacent to metropolitan counties, produced 52 percent of the dairy products in the country (Heimlich and Bernard, 1993). In 2007, Jakarta produced an estimated 80 000 tonnes of poultry meat and 400 tonnes of eggs within city limits (ICASEPS, 2008), with over 200 collection points and over 1 000 small slaughter facilities in the city. Most of the rest of the city's

---

[3] Self-reliance is related to self-sufficiency but not identical. Self-sufficiency implies producing all of one's own food while self-reliance means relying on one's own resources to obtain food.

BOX 8
FOOD AND BIOGAS PRODUCTION IN SHANGHAI

Shanghai follows China's strategy for food self sufficiency of mega-cities (Yi-Zhong and Zhangen, 2000). The total area of Shanghai covers 6 340.5 km², of which 13 percent is urban and the rest rural. The average population density within the Shanghai city limits is about 2 059 persons per km², very low compared with New York City (Manhattan), USA, which has 27 257 persons per km².

Agriculture contributes only 2 percent of the city's GDP, yet is a highly protected economy. About 8.5 million people in Shanghai have a job, 3.6 million of these in the agricultural production sector. The 2.7 million farmers represent 93 percent of the population of the rural parts of Shanghai, and 13 percent are full-time farmers (Yi-Zhong, and Zhangen, 2000). To prevent rapid turnover of agricultural to non-agricultural land, 80 percent of the arable land is protected under the Agricultural Protection Law. These measures have contributed to 100 percent and 90 percent respectively of the milk and eggs consumed in Shanghai being produced within the city limits. Local pork and poultry production cover just over half of the total supply to the city.

Peri-urban agriculture is encouraged to serve other functions besides food production. One of the most important is biogas production (Kangmin and Ho, 2006; Blobaum, 1980; Ru-Chen, 1981; Gan and Juan, 2008; IFAD, undated; Owens, 2007). According to the government's *Chinese Ecological White Paper* issued in 2002, the total amount of livestock and poultry wastes generated in the country reached 2.485 billion tonnes in 1995, about 3.9 times the total industrial solid wastes (Kangmin and Ho, 2006). Animal agricultural wastes are toxic pollutants when discharged into rivers and streams, but can be valuable resources if managed for compost or energy from methane. It is estimated that 10 million ha of farmland in China are seriously polluted by organic wastewater and solid wastes. China's national plan for biogas (Junfeng, 2007) calls for 4 700 large-scale biogas projects on livestock farms by 2010, thereby increasing biogas-using households by a further 31 million – to a total of 50 million or 35 percent of total rural households.

Source: Brinkley, 2010.

supply came from provinces within a two-hour drive. In the mid-1980s, up to 40 percent of the calories of urban dwellers in Kampala were provided by livestock raised in and close to the city (Smith and Olaloku, 1998). Shanghai produces almost all of the milk and eggs for its citizens within its city limits (see Box 8).

As cities expand and develop economically, animal production systems tend to move farther away. Residential areas encroach onto farmland, and as cropland moves outward, ruminant livestock moves outward in parallel to maintain proximity to available feed in the hinterlands. Pigs and poultry initially stay on the expanding fringes of the growing cities, but eventually are

encouraged to move further away to avoid environmental contamination (Gerber *et al.*, 2005; Costales *et al.*, 2006).

Cities also source food through international market chains, both formal and informal. Much of the official international trade in livestock products supplies urban populations. There is regular cross-border movement of live animals in Southeast Asia, Africa and parts of Latin America, although not all of it is recorded. The market chains that supply cities with poultry meat are defined by their diversity. Small- and medium-scale producers are located in peri-urban areas while large intensive production is located all over the world. At the same time, international market chains are both formal and informal. For example, a recent FAO study suggested that a million birds a month cross the border from China to Viet Nam informally.

There are no precise figures for the relative contribution of small- and large-scale production units to city food supplies. However, the worldwide trend is to upscale and concentrate. In the USA, the majority of production comes from large or very large units. In Brazil and Thailand, an increasing proportion of supply comes from large units, even though there are still many small-scale producers. In Viet Nam, where demand for livestock products has been growing steadily, avian influenza and other forces have pushed many small-scale producers out of business. Their market share was initially taken over by national companies but large regional players see this as an attractive domestic market opportunity and are gradually making inroads (McLeod and de Haan, 2009).

The structure of market chains that supply urban areas is changing. In some cases, markets within cities are being made more hygienic due to regulations, such as those in Hong Kong, Los Baños, in the Philippines, and Ho Chi Minh City, Viet Nam. In others, such as Jakarta, the smallest urban markets are being closed. Elsewhere, markets are changing their nature because of regulations. In Cairo, poultry are no longer assembled at physical markets but are traded through phone connections – when an order is placed, birds are moved from their production unit. This echoes the move towards a more virtual marketing system that followed the UK's FMD outbreak in 2001. Specialist local companies produce processed foods for the urban market within their own integrated chains, such as Farmers Choice in Kenya, which hires small-scale contract farmers to fatten pigs so that the company controls the source of meat for its own bacon, pork and sausages. Within cities, an increasing amount of product is sold in supermarkets (Reardon *et al.*, 2003; Reardon *et al.*, 2010). In the early stages of their development, supermarkets source products from a large variety of farms but, over time, they link to increasingly integrated chains.

The examples provided here demonstrate that there are many ways to define foodsheds and to provide sustainable food supplies to cities. The top-down policy measures used by China are very different from the American scheme to protect peri-urban agriculture through the coordinated efforts of private citizens and non-profits. The deliberate attempts in Brazil and Mexico to bring local food into cities differ from the more *laissez faire* approach in Nairobi that allows livestock to be brought within city limits and slaughtered there, even though this contravenes established regulations. As city populations grow, it will become increasingly important to discover and learn from successful examples.

PROSPECTS FOR LIVESTOCK FEEDING URBAN POPULATIONS

Urban populations are expected to continue to grow in numbers and proportion of the whole from the current 50 percent to 69 percent in 2050 (UN DESA/Population Division, World Urbanization Prospects, 2009). As stated by the UN Population Fund (UNFPA) in 2010, "[the] Urban population will grow to 4.9 billion by 2030. In comparison, the world's rural population is expected to *decrease* by some 28 million between 2005 and 2030. At the global level, *all* future population growth will thus be in towns and cities. ... The urban population of Africa and Asia is expected to double between 2000 and 2030. Meanwhile, the urban population of the developed world is expected to grow relatively little." UNFPA also stressed that the majority of new urban dwellers will be poor.

This presents a challenge to the livestock sector. As urban populations grow, there will be an increase in demand for some time, although the rate of growth will be limited by slow income growth in poor countries. The urban poor obtain much of their livestock source food from within or close to residential areas and it is reasonable to assume that they will continue to do so. However, there is a finite number of animals that can be kept within the residential area of a city, even when regulations are not applied to keep them out. Even in cities that have appropriate design and zoning to support peri-urban livestock, there is still a ceiling on what can be produced. To meet expanded demand, the area from which cities source their food is likely to become increasingly large.

Ruminants, which tend to be kept near feed supplies, may be located at quite some distance. This is not necessarily a problem for meat production, although transport economics will dictate the viable limits of foodsheds, and production units on average are likely to scale up. For dairy products, however, transport and processing logistics will dictate both the size of the "milkshed" and the scale of enterprises that can supply the city. In some places, it will continue to be viable to source milk through complex networks of small producers, as in India, while in much of Africa and Latin America, this will only be viable with investment in local cooling facilities and refrigerated transport or other methods of preservation.

Much of the growth in food demand is likely to be for poultry and pig products, and the need to keep food prices low will encourage continued up-scaling of these systems. However, large pig and poultry units concentrated around cities bring many problems such as disease risk, environmental pollution and animal welfare concerns. There are good reasons for their production units to be scattered, in order to avoid disease spread or the risk of financial disaster if there is an outbreak, and to be located in different places around the world where production economies are most promising.

Economic forces also may push large-scale livestock units away from densely populated areas, since land in these places is scarce and expensive. Studies in the USA have shown that farms in and near towns are generally smaller, produce more per hectare, have more diverse enterprises, and are more focused on high-value production than those further away (Heimlich and Bernard, 1993; Heimlich, 1988; Heimlich and Brooks, 1989). However, products such as vegetables and fruit, which can be marketed directly, have a greater price advantage in being produced close to town than most livestock products (Lopez *et al.*, 1988).

The most likely prospect is that there will be an ever-expanding series of production rings around large urban areas reflecting returns per unit of land, with the most productive and valuable crops (horticulture) and livestock (organic eggs and specialist animals) closest to cites, and larger commoditized units increasingly far away. The bulk of new city supplies of livestock products will need to come from intensive systems, because poor city dwellers need relatively cheap food, and this cannot be produced extensively in large quantities within reach of cities. Small-scale producers may find themselves unable to compete with prices or standards, particularly where they are few in number and have limited price negotiation power (Knips, 2006).

The points made previously about prospects for smallholder mixed farmers apply here as well. The opportunities for smallholder farmers to supply cities are specific to systems and certain countries, such as dairying in parts of the world where the informal market is strong, and production of small animals during the period when cities are expanding. Although the urban wealthy will be in the minority, they will still exist in sufficient numbers to exert substantial demand. They may continue to drive the demand for welfare and for local livestock breeds produced traditionally (Otte *et al.*, 2008). This presents an opportunity for some smallholder livestock producers to increase their income levels rather than lose out to industrial producers.

It is likely that large and very large units will increasingly predominate in feeding cities. However, intensive livestock will need to become much better at dealing with externalities from pollution, food safety hazards and zoonotic diseases, issues that are discussed in later chapters.

©FAO/PPLPI

Key points on three populations

The three populations examined in this section represent a continuum in the contribution of livestock to food security. Societies that depend on livestock, primarily grazing animals, for their most important source of livelihood and food security are shaped by the management of their livestock. Small-scale mixed farmers use livestock as part of a diverse livelihoods portfolio, seldom the main source of income or food but important because of their flexibility of use, asset value and ability to convert roughage and by-products into human-edible food. Urban populations, particularly those in large cities, are primarily consumers of livestock source foods that may be produced far away from the city.

LIVESTOCK-DEPENDENT SOCIETIES

Pastoralists and ranchers. Pastoralists, the largest number of livestock-dependent people at around 120 million, rely on their livestock

to provide food, income, transport and fuel. Ranchers, although fewer in number than pastoralists, make an important contribution to the supply of livestock products in their countries and the world through animals that they keep primarily as an income source. For both of these groups, animals convert human-inedible forage into human-edible protein and so contribute positively to the protein balance. By supporting their own population and generating surplus for export, livestock-dependent societies contribute to the world's supply of food as well as their own food access.

Systems under pressure. The global land area available for grazing is close to its biological limit for production under the prevailing climatic and soil fertility conditions, putting pastoralist systems under pressure. The area available for extensive grazing is unlikely to expand because of competition from agriculture and biofuel, human settlement and nature conservation. Decreased and more variable rainfall may require changes in management to cope with additional instability, while also creating new animal health challenges for these systems.

Investment and diversification. Existing

levels of production from livestock-dependent societies should be protected because of their contribution to the food supply and the protein balance. Investment in securing their access to markets is important as this offers livestock owners the opportunity to gain greater value from what they produce and to manage risk by managing stocking levels. The case of Mongolia illustrates that even highly livestock-dependent societies can be expected to become less dependent on livestock in the future. The current trend is a gradual movement of people into towns and away from pastoral agriculture. For those who choose to remain in rural areas, tourism, recreation and payment for environmental services such as wildlife conservation and carbon sequestration into grassland all offer complementary ways for livestock keepers to earn income.

SMALL-SCALE MIXED FARMERS

Integrated system. Livestock are a smaller part of the livelihoods portfolio for small-scale mixed farmers than they are in livestock-dependent societies, but they are still important. Livestock are managed as part of an integrated and tightly-woven system, in a way that fits the needs of the farm family, the available labour and the demands of other enterprises. Animals provide food, income, traction, manure, social capital, financial assets and a means of recycling crop wastes. They bring value, versatility and resilience to mixed farming households, which are more robust and food secure with animals than they would be without them.

Rural livelihoods. Small-scale mixed farms remain enormously important because of the large number of rural households they feed and provide with livelihoods. They also contribute to the food supply of developing countries and use and recycle resources effectively. Policies, public and private investments, and technology have supported small-scale dairying in India and parts of East Africa, where peri-urban small-scale dairy producers have good connections to milk markets and reasonable access to animal health services. However, most small-

scale farmers face limits to intensification, few have managed to upscale or specialize to a point where they can advance economically, and many depend partly on off-farm employment for their food security.

Limited potential. The case of Nepal illustrates both the benefits of livestock and the constraints faced by small-scale mixed farmers. Lack of opportunity or capital to increase farm sizes, limited assets and therefore limited access to credit, lack of investment capital, limited land availability, reduced access to communal land, higher unit costs than those of large producers, and restricted opportunities to market produce through physical distance or barriers imposed by quality and safety requirements are all factors that prevent many small-scale farmers in many places from expanding or intensifying their production.

Competition from large-scale producers. The supply of food to growing cities is an important growth areas in demand for livestock products, but here small-scale producers face strong competition from large-scale producers with intensive farms. Peri-urban small-scale farmers tend to be very successful at supplying urban populations in the early stages of demand growth, but less so as food safety and land use regulations become stricter. To compete successfully, they need to be credible competitors. For some, it is possible to become contract farmers to larger operations; for others, innovative approaches may offer the chance to access niche or specialist markets. For the remainder, especially in fast-growing developing countries, future prospects may be limited.

CITY POPULATIONS

Urban demand for livestock. Half the world's population lives in urban areas, and this proportion is estimated to increase to about 70 percent by mid-century. Urbanization has been associated with a rising demand for livestock products, for the most part because urban people are on average richer than rural dwellers. However, poor urban dwellers eat far less livestock source food than their richer counterparts and many are highly food insecure. Countries with growing urban populations as well as rising wealth must deal with two concurrent food security problems – a large proportion of the population that is undernourished but also a growing number of people who consume more than they need for health or have poorly balanced diets.

Feeding cities. The location of livestock production and the shape of livestock market chains are increasingly driven by the growth of cities. The cases of the USA, Kenya and China illustrate three approaches taken to feeding cities. Their national policies have been, respectively, a market-driven economy combined with strict land-use regulations, a *laissez faire* market economy with strong informal market chains, and a centrally-planned economy in which the objective is to have a high level of food self-sufficiency within tightly defined "foodsheds". While each has applied a different policy approach, all face the challenge of feeding expanding urban populations from what are likely to become increasingly large food supply areas.

Intensification issues. The need to keep food prices low for urban populations drives continued up-scaling and intensification of livestock, particularly of pig and poultry production. However, large livestock units concentrated around cities bring problems of disease risk, environmental pollution and animal welfare concerns. Intensive livestock will need to deal more effectively with externalities from pollution, food safety hazards and zoonotic diseases. Environmental regulations and the need to mitigate risk may also encourage production units to disperse, while economic forces tend to push large-scale livestock units away from densely populated areas where land is expensive.

Urban wealthy and smallholder opportunity. Although the urban wealthy are in the minority, they exist in sufficient numbers to exert substantial demand and will continue to do so as populations grow. They will continue to drive the demand for welfare and for local livestock breeds produced traditionally. This presents an opportunity for some smallholder livestock producers to increase their income levels rather than lose out to industrial producers.

©FAO/Giuseppe Bizzarri

Feeding the future

Producing enough food

We can safely assume two things about the next 40 years: the demand for livestock products will continue to grow, and it will become increasingly challenging to meet that demand. At some point, perhaps as soon as 2050, it is estimated that there will be 9.15 billion people to feed, 1.3 times as many as in 2010 (UN Population Division, 2009). Much of the new population will be urban (UNFPA, 2010). Based on estimates published in 2006, the expanded population is expected to consume almost twice as much animal protein as today. While the projections are for a lower annual rate of growth than occurred during the livestock revolution, doubling supply would still place a considerable burden on already strained natural resources. This, in turn, would drive up the prices of livestock products and threaten food access by the poor.

However, there is a great deal of waste in food systems. Natural resources are not always converted efficiently into meat, milk or eggs, and a great deal of the food currently produced does not reach the plate. Improving efficiency and minimizing waste throughout livestock value chains could go a long way towards meeting increased demand. This chapter reviews the assumptions on which the projected demand for food is based and discusses how accurate they are likely to be. It then examines the three main systems in which livestock source food is produced to identify where efficiency might be improved and waste reduced.

HOW MUCH LIVESTOCK SOURCE FOOD WILL BE NEEDED?

The most complete published projections at the time of writing (FAO, 2006c) suggest that in 2050, 2.3 times as much poultry meat and between 1.4 and 1.8 times as much of other livestock products will be consumed as in 2010 (Table 16). The additional demand beyond that expected from population growth will result from increases in income encouraging a higher consumption per person. The largest growth is expected in developing countries, which are anticipated to overtake developed countries in their total consumption of livestock products. The figures in Table 16 assume that purchas-

TABLE 16

PROJECTED TOTAL CONSUMPTION OF MEAT AND DAIRY PRODUCTS

	2010	2020	2030	2050	2050/2010
		(million tonnes)			
WORLD					
All meat	268.7	319.3	380.8	463.8	173%
Bovine meat	67.3	77.3	88.9	106.3	158%
Ovine meat	13.2	15.7	18.5	23.5	178%
Pig meat	102.3	115.3	129.9	140.7	137%
Poultry meat	85.9	111.0	143.5	193.3	225%
Dairy not butter	657.3	755.4	868.1	1 038.4	158%
DEVELOPING COUNTRIES					
All meat	158.3	200.8	256.1	330.4	209%
Bovine meat	35.1	43.6	54.2	70.2	200%
Ovine meat	10.1	12.5	15.6	20.6	204%
Pig meat	62.8	74.3	88.0	99.2	158%
Poultry meat	50.4	70.4	98.3	140.4	279%
Dairy not butter	296.2	379.2	485.3	640.9	216%

PERCENT OF TOTAL CONSUMPTION IN DEVELOPING COUNTRIES

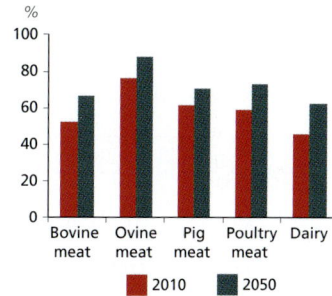

Source: FAO, 2006c. Some calculations by authors.
Note these figures are based on World Population Prospects: The 2002 Revision.

ing power and eating habits will follow patterns broadly similar to those recorded in recent years. As changes in any of these drivers could change the projections, each of them will be examined in turn, starting with the population estimates.

Population estimates. In 2002, the UN projected a population of 6.83 billion in 2010 and 8.91 billion in 2050 with a peak population of 9.2 billion, possibly in 2075. In 2008, the figures for 2010 to 2050 were revised slightly upwards, as shown in Table 17. However, the growth between 2010 and 2050 is virtually identical in both estimates, at 130 percent and 132 percent respectively. Using the new population estimates, the total demand for livestock products might be expected to increase slightly, but the growth between 2010 and 2050 should be very close to what is shown in Table 16. On the basis of population growth, therefore, it seems reasonable to use the current projections of demand for livestock products. The assumptions related to purchasing power of livestock products will be examined next.

Consumption growth. The projected growth in consumption per person, shown in Table 18, is based on total consumption figures from Table 16 and the 2002 population estimates on which those projections were based. The 2007–08 economic crisis temporarily reduced the growth rate of GDP and therefore the purchasing power for livestock products, but expectations are that the effect will not be prolonged and that average long-term growth will be as expected.

Production costs. Purchasing power is also affected by the price of livestock products, which in turn is affected by the cost of production. The latter could increase if feed and fuel energy become more expensive, water becomes scarcer or livestock value chains are increasingly required to bear the costs of the negative externalities they create. All of these are possible. Crops that can be used as both food and feed are likely to increase in price (Thornton, 2010), since increased yields will depend in part on fossil fuels and scarce minerals. Competition for bioenergy also may drive up prices, although new technol-

TABLE 17

PROJECTED HUMAN POPULATION FROM 2002 AND 2008 ESTIMATES

	2010	2020	2030	2050	GROWTH 2010 TO 2050
	(population billions)				
2002 projections	6.83	7.54	8.13	8.91	130%
2008 projections	6.91	7.67	8.31	9.15	132%

Sources: World Population Prospects 2002 and 2008.

TABLE 18

PROJECTED CONSUMPTION OF LIVESTOCK PRODUCTS PER BILLION PEOPLE BASED ON 2002 POPULATION ESTIMATES

	2010	2020	2030	2050	GROWTH 2010 TO 2050
Human population billions	6.83	7.54	8.13	8.91	
	(Consumption million tonnes per billion people)				
Bovine meat	9.85	10.25	10.93	11.93	121%
Ovine meat	1.94	2.08	2.28	2.64	136%
Pig meat	14.98	15.29	15.98	15.79	105%
Poultry meat	12.58	14.72	17.65	21.69	173%
Dairy	96.24	100.19	106.77	116.55	121%

Sources: FAO, 2006c; World Population Prospects, 2002. Some calculations by authors.

ogy is likely to make it possible to use a wider range of non-food inputs to produce biofuel. Water availability is also a serious consideration, since the proportion of people living in water-stressed regions is expected to rise to 64 percent in 2025 compared to 38 percent in 2002 (Rosegrant *et al.*, 2002) and livestock are a major user of fresh water, currently estimated at 20 percent of green water flow[4] (Deutsch *et al.*, 2010). Livestock production creates externalities through water pollution and emission of greenhouse gases – costs for which it does not currently have to account. Research and pilot projects are exploring the extent to which environmental

services provided by livestock, such as soil carbon sequestration through grazing land management (Conant and Paustian, 2002; Conant, 2010; Henderson *et al.*, in press), as well as more efficient recycling practices such as biogas production, could mitigate environmental problems and associated costs.

Combining all of these factors, there is a strong possibility that prices of livestock products will increase. Projections by OECD and FAO suggest that average prices of poultry meat and beef will be higher in real terms during 2010–19 than they were in 1997–2006, with limits in supply, higher feed costs and rising demand all contributing to the effect (OECD-FAO, 2010). Average dairy prices in real terms are expected to be 16–45 percent higher in 2010–19 compared to 1997–2006. If this happens, it could reduce

---

[4] Green water is the precipitation on land that is stored in the soil or temporarily stays on top of the soil or vegetation. It is the source from which crops draw their water.

accessibility particularly for poor urban dwellers and result in a change in diet for the less well-off, including more vegetable protein and cheaper cuts of meat. The possibilities for improved technology to increase productivity are discussed in the next section.

Price of livestock protein. The relative price of livestock protein and substitute proteins also affects the demand for livestock products. The biggest direct competitor is fish, which is estimated to provide 22 percent of the protein intake in sub-Saharan Africa (FAO, 2006d) and 50 percent or more in some small island developing states and some ten other countries (FAO, 2008c). In the past 20 years, fish consumption per person has remained fairly stable (FAO, 2008c) while consumption of livestock products has grown, but this could change if relative prices change.

With marine stocks dwindling and caught sea fish more expensive, sea and inland aquaculture have become more important. Marine aquaculture production grew from 16.4 to 20.1 billion tonnes between 2002 and 2006, and inland aquaculture from 24 to 31.6 billion tonnes during the same period (FAO, 2008c) with two-thirds of all production in China. Aquaculture is now estimated to be responsible for almost 50 percent of fish consumption and it is set to overtake capture fisheries as a source of food fish (FAO, 2010b).

Some farmed fish are highly efficient feed converters of the same feeds used for livestock (fishmeal, soya and cereals), take little space and, in some cases, do not require fresh water. There are problems associated with intensive rearing such as contamination of the marine environment with algae, over-use of antibiotics, over-fishing to provide low-value catch fish as feed, and contamination of fish with toxic chemicals. If these can be solved (Black, 2001; Stokstad, 2004), farmed fish have the potential to take a larger share of protein consumption.

Insects caught in the wild are consumed by over 2 billion people in Latin America, Asia and Oceania (FAO, undated), contributing to food supply and to the livelihoods of those who harvest them. Edible insects have the potential to be "farmed" and recent research suggests that they could be more efficient and produce lower methane emission than livestock (Oonincx *et al.*, 2010).

Meat produced "*in vitro*" (artificially) offers a possible future competitor to meat from animals for those who wish to consume meat sustainably or have concerns about animal welfare. It has the potential advantages of using less water and energy and being more welfare-friendly than rearing animals, but the technology has some way to go before it can produce marketable meat. Current techniques involve growing cultures from stem cells of farm animals into 3-dimensional muscle structures. Stem cells are currently obtained from muscle removed by biopsy and multiplied in culture, although it may in time be possible to maintain an independent stock of stem cells.

It is difficult to bulk up the cells, as each cell only divides a certain number of times (Jones, 2010), and while growth media not containing animal products are available, they are expensive. The resulting meat has poor texture and will need to have fat cells grown together with the muscle to improve its taste as well as added micronutrients before it is viable as a meat substitute. It is also expensive to produce, costing between €3 300 per tonne and €3 500 per tonne (The *In Vitro* Meat Consortium, 2008). However, this is a relatively new technology with relatively little spent on research thus far. Within the next 40 years, it may well become a part of the diet for some consumers.

Consumer lifestyle. Voluntary lifestyle choices, particularly by wealthier consumers, could result in consumption of fewer livestock products, particularly red meat. The newly wealthy have tended to eat more livestock products, particularly red meat and fatty foods, while some of the established wealthy tend to gradually diversify their dietary habits towards different cuisines and sources, "green" products and healthier diets. The current projections take these trends into account to some extent. McMichael *et al.* (2007) suggest that the average global consumption of meat should be approximately 90 g a day, compared with the current 100 g, and that not more than 50 g should come from red meat from ruminants. If this target were achieved, it would lower the peak demand for meat. However, government-sponsored nutritional and healthy-eating programmes have had limited success in changing dietary preference. It may be possible to envisage policies that could reduce over-consumption of meat through taxes and legislation, but it is impossible to imagine any economic incentive or legislative process that would not restrict access by poor consumers, who would benefit nutritionally from consuming animal products of high quality. Therefore any changes to diet are likely to be driven primarily through education, choice and exposure to healthy food. Strategies to bring healthy food within closer reach of everyone in the urban community could be helpful in this regard. In the UK, it is not the government alone but coalitions of the public and private sector that are driving current changes in consumption (Harding, 2010).

Pulling together all of the factors mentioned here, it seems likely that FAO's 2006 consumption projections represent a ceiling. Demographic and economic trends may act to keep livestock consumption at the forecast levels, while production costs and competition particularly from fish are likely to dampen consumption growth for livestock products. For the time being, it seems wise to assume that the demand for meat may grow by as much as 1.7 times and for milk by 1.6 times, as projected, and to consider whether it is feasible to produce that much.

REDUCING WASTE

The growth in production that took place during the livestock revolution was largely a result of an increase in the number of animals. Demand grew so fast that it was difficult for productivity improvements to keep up. Now, it is hard to envisage meeting projected demand by keeping twice as many poultry, 80 percent more small ruminants, 50 percent more cattle and 40 percent more pigs, using the same level of natural resources that they currently use. Part of any increase will need to be driven by efforts to convert more of the existing natural resources into food on the plate. In other words, efficiency needs to increase or, looking from another angle, there is a need to reduce waste of natural resources. In both cases, the end point is the same, but focussing on waste puts a spotlight on what is thrown away and might be recycled.

Waste occurs throughout livestock food systems. It can be due to production inefficiency resulting from disease or poor feeding. It also can result from loss of food between production and the plate, which may amount to as much as 33 percent for all global food production (Stuart, 2009). Food lost at or near the point of consumption, because of food safety and quality requirements, is a problem, but it will not be addressed here because there is little that the livestock sector can do about it. Losses that occur on the farm or in marketing and primary processing of livestock commodities are within the influence of the livestock sector and therefore will receive more attention.

Two issues related to waste reduction can be assessed further.

Choice of livestock system. If a larger percent of the world's livestock protein were produced within grazing and low-intensity mixed systems, would this leave more plant protein to be eaten by humans? According to FAO (2009b), the reality is not that simple. The main problem of food security is not currently one of supply but of demand. The 925 million undernourished people are not undernourished because the global food supply is deficient, but because they cannot afford to buy food or they live in places or societies where it is hard to obtain. Reducing the grain fed to livestock would not ensure that these people could access food. Neither would it automatically result in more plant protein being grown, as it might reduce the prices for those commodities to a level where it would be less attractive to grow them, although the higher number of people to be fed and increasing resource pressure may change this in future. Intensive systems also have economies of scale that make it possible to produce livestock protein in large quantities relatively cheaply, an important consideration for growing urban populations. The less intensive systems are an excellent option to supply food to rural populations with access to short food chains, or to consumers who can afford to buy "green" products, but they are less practical for the majority of city populations.

Livestock and waste recycling. Livestock have a role as recyclers of waste. Mixed farming systems are known to be particularly good at this, but even intensive production systems use by-products. For example, distiller's dried grains with solubles (DDGSs), a by-product of biofuel production, can substitute for grain in animal feed, particularly dairy and beef. In doing so, it contributes to the food balance and helps improve the economic viability of biofuel production. Intensive livestock also can use other industrial by-products, including some from the food industry, provided they are processed appropriately.

Inefficiencies and waste arise in different ways and locations in the three food systems discussed in earlier chapters. We therefore return to the three food security situations – livestock-dependent societies, small-scale mixed farmers, and city dwellers – with their associated livestock production and marketing chains, to examine critical areas of inefficiency for each situation and to suggest where the emphasis might lie in addressing the inefficiencies.

LIVESTOCK-DEPENDENT SOCIETIES

The pastoralist and ranching systems associated with livestock dependent societies are well adapted to their environments and quite efficient at using the forage they are able to access. Survival of animals is as much a yardstick of efficiency as production per animal, and tradi-

tional systems as well as ranching adopt forage management and conservation systems that will take animals through severe winters and dry seasons. In the future, the environmental restrictions on these systems are likely to persist or even worsen. Thornton and Gerber (2010) identify droughts, floods, temperature stress and reduced water availability as serious problems for grazing systems – events that are difficult to predict and even more difficult to mitigate. The following identifies areas that have possibilities for improvement.

Pasture management. Pasture restoration or, even better, good management that keeps pastures from being degraded in the first place and avoids the waste and high cost of restoration, offer the possibility of sequestering carbon and mitigating greenhouse gas emissions (Thornton and Herrero, 2010; Conant, 2010). Unfortunately, pasture degradation seems hard to prevent, particularly in pastoralist areas where institutions for resource management are weak. In addition to well known problems associated with loss of land to agriculture and decisions by herders to overstock, the impacts of climate change are adding extra disruption.

Animal health. Disease is an enormous source of inefficiency and waste. Diseases such as *peste des petits ruminants*, contagious bovine and caprine pleuropneumonia, swine fevers and some tick borne-diseases can kill animals that have been reared for months or years before they are fully productive, while internal parasites, tick damage, foot-and-mouth disease and abortions caused by brucellosis can reduce their ability to grow or produce milk. Zoonotic diseases which are passed from animals to people, such as brucellosis and tuberculosis, reduce the ability of people to benefit from their food.

Reaching livestock-dependent societies with well organized vaccination campaigns and essential drugs is critical to prevent production waste. This is logistically possible but institutionally challenging with problems in both supply and

©FAO/Giuseppe Bizzarri

demand. During the Pan African Rinderpest Campaign, thousands of cattle were vaccinated annually, even in the most remote areas. This had a parallel benefit for sheep and goat owners whose animals could be vaccinated against other diseases at the same time but, when donor funds were withdrawn, the service stopped. Even when a supply chain for drugs and vaccines goes to every small town, providing ready access for livestock owners, many choose not to vaccinate their animals routinely, particularly the small animals of lower value. There is also limited quality control over drugs and vaccines that are sold in remote areas (Ngutua *et al.*, undated; Leyland and Akwabai, undated), and many local sales merchants do not have suitable cold storage to keep the products in good condition.

Governments often perceive that the cost of maintaining an animal health service in remote areas is too high. Ranchers pay for private veterinary services but these services are often completely absent for pastoralists. If global demand for livestock food outstripped supply and the value of products coming from livestock-dependent societies increased, there could be a strong incentive to invest in animal health to prevent waste. Alternatively, investment in cost-sharing systems, where farmers and the government each contribute, could prove viable in some places (Mission East, 2010). Para-professional

veterinary services of various kinds have been tried and have been partly successful, but will need to be more sustainably supported in a variety of forms to have a long-term impact on reducing the waste caused by animal health problems.

Transportation infrastructure. Losses occur in marketing because of the long distances that animals and products must be transported. Poor roads and often the need to pass through conflict areas make it hard to provide reliable transportation. Animals travelling in poorly designed lorries without adequate water lose weight, suffer dehydration and bruising, and may die. Milk is in danger of spoilage unless local coolers and refrigerated trucks are available. If prices are low or transport unavailable, any excess milk that cannot be consumed by calves or people will be wasted. There are technical solutions to these problems when a demand exists for the product. Milk coolers and alternative forms of preservation such as lactoperoxidase have been provided in remote places in Africa (FAO, 2005), rest stops have been built where animals can be given water, and lorries are available that improve animal welfare during transport. The challenge, as always, is to find funds to invest in the necessary infrastructure and technology.

Markets. From a food security perspective, an emphasis on markets is critical for livestock-dependent societies. Ranchers and governments in developed countries are very well aware of this. In pastoralist systems, innovative approaches to improving access to markets for live animals and livestock products are essential and so are programmes to pay for environmental services. Together, these can be an incentive to reduce production and transport losses, and provide livestock-dependent communities with the means to co-finance animal health, pasture management and better transport facilities.

SMALL-SCALE MIXED FARMERS
Small-scale mixed farmers are efficient at using and recycling natural resources. Their animals

eat crop residues, kitchen scraps, snails and insects. They grow forage at the edge of crop fields or around houses, or cut and carry it from communal grazing areas, forests or the side of the road. Mixed farming is probably the most environmentally benign agricultural production system and it has a great deal to contribute to minimizing waste, especially with all of the opportunities it offers for nutrient recycling (LEAD, undated). Given the number of small-scale mixed farms, if most of them increased their efficiency by even a small amount, it would be beneficial for the global food supply and food security. However, there are currently three major sources of waste that need to be addressed.

Poor animal health. Animals on small-scale mixed farms have a high prevalence of "production" diseases such as external and internal parasites (Mukhebi, 1996; Over *et al.*, 1992) and mastitis (TECA, undated; Byarugaba *et al.*, 2008) that rarely cause death but always reduce performance (Tisdell *et al.*, 1999), as well as zoonotic diseases such as brucellosis and TB that cause human illness and production losses. These can generally be controlled if farmers invest in basic prevention measures. Understandably, they tend to do this more for higher value animals such as dairy cows. Farmer cooperatives have proved valuable for small-scale dairy farmers to obtain animal health inputs, as have projects that give or loan animals to these farmers but require them to provide certain standards of housing and care.

Poor feeding. Poor feeding is problematic on its own, but even more so when combined with animal health problems. When traditional livestock breeds are reared in research stations, fed a balanced diet and provided with health care, they perform credibly compared to exotic breeds (Mhlanga *et al.*, 1999) and can out-perform those on mixed farms. Although a great deal of research has been done over the years on feeding animals in mixed farming systems, and some crop-breeding programmes have im-

proved the quality of stover (stems), the problem of feed shortage still persists. Recent work with small-scale dairy farmers in Ethiopia found that they prioritized lack of feed over disease problems (K. De Balogh, FAO, pers. comm., based on unpublished research). Since one of the major constraints for intensification of small-scale livestock production is the lack of good quality feed resources, it will be worth persisting with research into ways to improve use of locally available feed resources, especially those not competing with human food. There may be long-term potential to breed for improved ability to digest cellulose (National Research Council, 2009). In Anand, India, through the efforts of the National Dairy Development Board (NDDB), milk production has been increased sustainably by feeding diets containing cereal straws, roughages and oilseed cakes. In Africa, 427 million tonnes of cereal residues (based on FAOSTAT grain data and average ratios of grain to residues) and 9.2 million tonnes of oilseed cake are available annually (FAOSTAT), but there is a logistical challenge to making it accessible. Exports of oilseed cake can be a strong competitor to domestic uses, but oilseed cake is produced in plants processing the primary products where it is not always easily accessible to small-scale farmers.

Post-harvest losses. A third source of loss is post-harvest spoilage of products. Stuart (2009) suggests that more of the loss occurs at the retail end of the chain in developed countries, while in developing countries more is lost on-farm. Spoilage on the farm is a particular concern for dairy farmers, and a great deal of effort has gone into finding small-scale technology for preserving milk (FAO, 2005). Meanwhile, Indian dairy farmers in several states benefit from daily or twice daily collection of their milk.

As previously discussed, marketing their products is a common constraint for small-scale mixed farmers. While for livestock-dependent societies, the challenge is mainly one of distance to markets, small-scale mixed farmers face problems of barriers posed by food safety and quality demands and of a concentrated market chain that makes it difficult for them to compete. The importance of this in the context of waste is that without a market outlet, farmers have little incentive to experiment with new technology that will make them more efficient. Food quality and safety regulations can contribute to minimizing losses further along the chain, by reducing waste at slaughterhouses and retail points. Supporting small-scale mixed farmers in improving their quality standards and biosecurity, while at the same time continuing to recycle waste efficiently, would be a very positive contribution to food security for the future. Not every small farmer will be able to benefit but for some, traditional products certified as safe or from a valued production system have the potential to command a higher price and attract investment into marketing.

FEEDING CITIES FROM LARGE-SCALE INTENSIVE PRODUCTION

Much of the future demand for livestock products, particularly for urban populations, will have to be met by integrated value chains served by intensive medium- and large-scale production units with the potential to increase production per animal, per unit of land and per unit of time. These food systems are economically competitive but can be highly wasteful of natural resources. However, they do have the potential to improve.

A large part of the loss is at the retail end of the value chain, to meet the demands placed on supermarkets and fast-food retailers for quality and freshness (Stuart, 2009). Feeding waste food to animals is severely restricted in developed countries because of concerns about the safety and variable quality of the waste (Kawashima, 2002). While livestock source food is not safe to feed to animals unless very thoroughly processed, because of the risk of disease spread, there are other examples of animals being used to recycle other kinds of organic waste. One scheme recycled 30 000 tonnes of waste a year from the USA city of Philadelphia through pigs owned by a cooperative in New Jersey. This was

an estimated 8 to 10 percent of Philadelphia's municipal waste (Maykuth, 1998).

Food safety crises are frequent causes of waste in developed country food chains, examples being the 2009 withdrawal of ground beef from California markets because of e-coli contamination, the 2010 contamination of milk products by melamine in China and the 2011 contamination of eggs by dioxin in eggs in Germany. There is constant upgrading of safety management throughout food chains but since consumers and retailers pursue a near-zero risk policy, this kind of waste will always exist to some extent.

Moving further down the chain, there is waste during slaughter and processing. Some of this is due to parts of the animal or whole carcases being condemned or downgraded for health reasons or bruising (Martinez *et al.*, 2007; Tiong and Bing, 1989). Investment in animal health and welfare can prevent some of these losses.

At the farm, greater use of the agro-industrial by-products that make up part of animal feed could reduce the amount of human-edible food fed to livestock. Intensive livestock in the emerging economies make quite effective use of agro-industrial by-products. For example, in India's poultry industry, feed manufacturers include waste from the food industry, the gum and starch industry, fruit and vegetable processing and the alcohol industry in poultry feed (Balakrishnan, 2002). This forms quite a large proportion of India's feed input (H. Steinfeld, pers. comm. based on recent unpublished analysis), while the Malaysian ruminant industry uses crop residues and food industry by-products in ruminant feeds. However, there are very strict restrictions on the use of the livestock industry's own by-products. For example, meat and bone meal is forbidden to be used in animal feed because of its potential to spread BSE. In the UK, approximately 60 000 tonnes annually of ash from incineration of meat and bone meal is sent to landfill (Environment Agency, UK, undated).

Feeding and health systems are also important to exploit the genetic potential for feed conversion. Therefore another way to limit waste is to ensure that all farmers move closer to the standards set by the most productive. Ruminant systems still have some potential to increase their productivity through breeding (Thornton, 2010), particularly if the balance of grain to roughage can be reduced. Some would argue that feedlot cattle are fed too much grain for their own health or for optimum productivity. Animal welfare standards, which are becoming more demanding in developed countries, may increasingly influence the limits on feed conversion and other productivity improvements. For example, there will be no battery production of eggs in the EU after 2015, and the use of bovine somatotrophin has been banned there for several years.

It is possible to recycle livestock waste through large-scale anaerobic digesters that turn solid food waste into biogas, or large-scale composters to turn food waste into compost that can then be used as farm fertilizer (Harvey, 2010). China has emphasized biogas production and some European countries are placing emphasis on using biogas technology (Kaiser, undated).

In addition to feed conversion, indicators that measure the environmental impact of production are also important, because this affects the quality of natural resources on which production depends. Manure from pig and dairy enterprises contributes to greenhouse gas emissions through the handling and storage of slurry (Henderson *et al.*, in press), but this can be processed through biogas units. Manure from grazing livestock creates N_2O emissions when it is broken down by microbes (Steinfeld *et al.*, 2006). Beef is the most emission-intensive meat while chicken is the least (Fiala, 2008). Improved productivity, on the whole, reduces emissions per unit of meat produced.

There is quite a strong potential to reduce waste throughout the food systems that supply livestock source food to cities. At each point on the chain, technology is either available or being investigated that could be helpful in this regard. In both developed and emerging economies, the private sector is making quite substantial

©FAO/Olivier Thuillier

investments in technology that reduces waste and saves costs. The role of the public sector is to provide an environment in which there is an incentive to minimize waste throughout the market.

However, this does require a balancing act among welfare (which may indicate less intensive farming), productivity (more intensive farming), emission reduction (less beef) and safety (certified biosecure farming and no recycling of animal products through livestock). Middle class consumers have not yet begun to take an interest in waste from livestock systems. Once they do, it may lead to a small overall reduction in the demand for animal products, and a small shift in demand towards food products with waste-saving credentials.

This chapter has thrown up several challenges for the livestock sector, and some possible directions, such as efforts to minimize waste and increase efficiency, that will contribute to assuring livestock's role in food security for the future. The next chapter looks at possible directions for building resilience into a sector that is experiencing the growing pains of increased demand in a globalizing world that brings with it new threats of disease and external economic shocks as well as those caused by more extreme weather events linked to climate change.

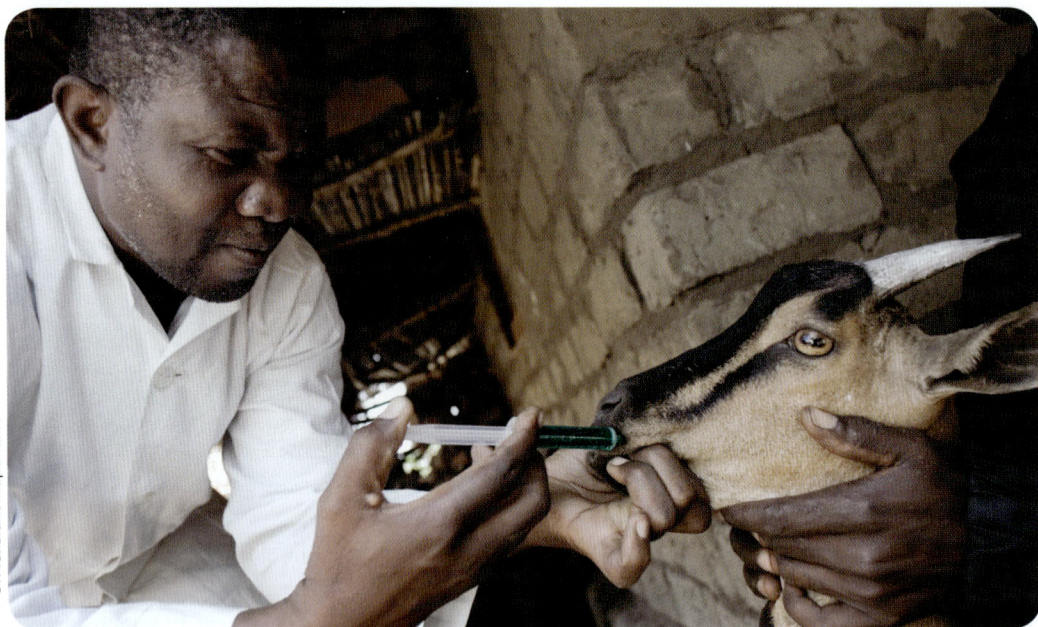

©FAO/Giulio Napolitano

Building resilience

The livestock revolution was characterized by rapid increases in production, driven by rising livestock populations and income on the demand side and cheap feed and fuel on the supply side. Today, demand continues to grow in spite of economic shocks, but supply conditions have changed – a scenario that has profound implications for the way the livestock sector will develop and the role it will play in food security in the future. As the previous chapter discussed, the pressures on natural resources may force the price of livestock source foods to rise, making them less accessible to the poor, but it also proposed that improving efficiency and reducing waste in livestock production will make important contributions to ensuring the supply and accessibility of livestock source food.

Today's livestock sector must be prepared to respond with a shift in focus and investment towards building greater resilience into food systems, meaning an increased ability to deal with change and recover from shocks. There is

increasing concern about the instability of food supply and access in what are termed "protracted crises" (FAO, 2010a). This chapter therefore reviews some of the factors that may create vulnerability in livestock food systems and looks at ways in which they can be mitigated.

Livestock have a certain inherent resilience as ruminants and camelids can withstand a wide range of temperature and moisture conditions while poultry and pigs are less adaptable to heat and cold but can easily be housed. Notwithstanding the adaptability of animals, however, livestock food systems face hazards from several sources. Climate change is creating new shocks and trends; boths are certain but hard to predict and have potential to make the production environment uncertain in ways similar to El Niño events. It also will probably create future hotspots, with higher temperatures and lower rainfall which will affect water availability and average temperatures, both critical to crop production.

The following section looks at three potential hazards the livestock sector faces: water shortages, spread of persistent or emergence of new diseases including those transmissible to humans, and market volatility, particularly for producers trying to import feed or export food, and for food-importing countries and cities.

WATER SHORTAGE

With an increasingly large population living under conditions of water stress (Rosegrant *et al.*, 2002), agricultural systems will need to develop more built-in resilience particularly related to water use, and some crops may need to be relocated or different ones grown. Irrigated crops occupied around 20 percent of the arable area in 2002, an increase from 16 percent in 1980, but there were large regional differences.

In sub-Saharan Africa, only 4 percent of arable land was irrigated in 2002 compared with 42 percent in South Asia (FAO, 2008b). In the future, irrigated cropland may need to expand if larger areas become water stressed, but this irrigated agriculture will only be viable if it is highly efficient and more proficient at using water and preventing pollution from runoff than much of today's production. Steinfeld *et al.* (2010) identified a number of policy instruments which reflect scarcity, such as water pricing, pollution taxes and state recovery of the maintenance costs of irrigation systems. The success of some water-scarce places, such as Israel, shows how much can be done by careful use of water and recycling water resources.

Livestock systems are affected by water and temperature ranges but, in addition to direct changes in response to climate change, they can be expected to undergo second order changes that follow the shifts in agriculture.

- Grazing systems. The location of grazing and browsing livestock has always been determined in relation to crops, with livestock taking the land that is too wet, dry, mountainous, distant or stony for cultivation.
- Intensive systems. Animals on feedlots tend to be located near the source of crops or agro-industrial by-products. Intensive pigs and poultry have more flexibility and, since their feed is brought to them, they provide high returns on each unit of land and can be located quite close to urban areas. They

also have the potential to relocate to areas that are marginal for crops, perhaps to the fringes of deserts where solar powered air-conditioning and pumps for waste may provide a solution to rising energy costs. However "landless" livestock (those that are housed and take up little physical space) are major users of water through their feed, which means the efficiencies in crop water use will factor into livestock systems.

Although the livestock sector is in some sense a secondary responder to water shortage problems – due to its responding to changes in cropping systems – it can also take positive actions to deal with pressures on water stress. In livestock-dependent societies, pasture improvement can help livestock keepers adapt to climate change, and changes in land tenure may also be necessary to provide pastoralists the incentive to make necessary investments (Steinfeld *et al.*, 2010).

Cropland for food crops is already becoming squeezed by growing civil and industrial infrastructure, biofuel needs and nature conservation. If it must also be farmed differently to conserve water, there may be even less left for livestock. More than ever, animals will need to fit into the gaps left by cropping, using residues and roughage, wasting as little as possible of scarce inputs and having the flexibility to cope with fluctuations in crop yields. It may be necessary to rediscover crops suitable for small-scale mixed farming systems so that more of their by-products are available for livestock. Interactions between livestock and crops, lost when systems scaled up and intensified, may need to be revisited, not simply by returning to the past but by thinking innovatively about what is possible with the systems of the future. All of this is a far cry from the early days of the livestock revolution when feed appeared to be in limitless supply.

HUMAN AND ANIMAL HEALTH THREATS

Sudden disease shocks are problematic for food supplies. Persistent diseases such as internal and external parasites or mastitis create vulnerability by eroding the production and income base of livestock keepers. Certain animal and human diseases are likely to expand their range as a result of climate change, especially when they or their vectors (insects, mites and ticks) depend on warm annual temperatures and humidity. In new ecological niches, they will undoubtedly find new hosts to infect. Concern for new threats to human health is setting the direction for the major animal and public health initiatives of the international community, translating into various efforts in support of "One Health" and related initiatives (FAO/OIE/WHO/UNSIC/UNICEF/World Bank, 2008; Public Health of Canada, 2009; CDC, 2010).

To mitigate the risk of diseases, the focus of the animal health system will need to change. Currently, the attention of animal health professionals and finance systems is focussed on preventing the *transmission* of diseases when outbreaks occur, and the *prevention* of disease through import restrictions, quarantine and screening, biosecurity measures, and damping down the impact and spread of diseases using vaccination when it is available. Intervention measures to break transmission and prevention are important, but for the food systems of the future it will not be enough to focus only on them. Neither confronts the *root causes of disease emergence* and, as a result, veterinary and public health systems are constantly running to catch up with diseases that represent a threat to the stability of food supplies and to human health.

To build sustainability and resilience, more attention is needed to the drivers of disease. These fall into three areas, described in Box 9, each of which relates to a different kind of disease threat, creates a different kind of impact and therefore requires a different kind of response.

Building animal health and veterinary public health systems from knowledge of the drivers will make it possible for them to be more proactive in supporting food production. Food security is an important concern to the international animal health community but arguably a secondary one to the concern of dealing with dis-

ease. However, well managed disease control initiatives can minimize the market shocks caused by livestock diseases or their control. This is translating into research on drivers of disease, with more detailed contingency planning and business response planning in developed countries, and increased investment in response capacity and biosecurity in developing countries. If successful these various initiatives would improve the stability of food supplies, but there are still major institutional and investment gaps to be filled (Perry and Sones, 2008; McLeod and Honhold, 2010).

VOLATILE MARKETS FOR FEED AND LIVESTOCK PRODUCTS

Farmers no longer can rely on cheap feed. Prices have risen since the height of the livestock revolution and, equally important, they are unpredictable (Von Braun, 2008; Walker, 2010; BFREPA, 2010; *Beef Magazine*, 2008). The cost of fuel, competition from human food, biofuel and aquaculture, and climate shocks all contribute to these effects.

Market volatility for livestock products can occur because of disease shocks, other natural disasters, natural price cycles and economic shocks that reduce consumption. Longer term market changes occur when changes are made to production systems to improve biosecurity, which often result in smallholders being excluded. As discussed in earlier chapters, small-scale producers and pastoralists at the end of long market chains are particularly vulnerable as they have very little control over the market. Some efforts can be made to connect them to more lucrative markets (e.g. contract farming, cooperative action, niche markets) and to exclude them from some of the shock effects (e.g. commodity trading rather than disease free zones), but they remain vulnerable to competition from larger players. Large producers and companies are also vulnerable because of the size of the asset invested but big companies have some potential to diversify into feed, drugs, more than one species of livestock, or processed as well as fresh

BOX 9
DRIVERS OF DISEASE AND POSSIBLE RESPONSES

Land use. Big changes in the patterns of land use have been driven by climate change, urbanization and global movements of people in response to opportunities or crises. This allows disease agents to move to new geographic areas with similar ecosystems, adapt and survive. Disease agents on the move cause food instability when they initially infect naive animal populations. When a disease problem is caused by land-use changes and human demographic factors, it may not be possible to prevent its moving into a new environment, but early knowledge of a new problem makes it possible to take steps to protect animals by promoting vaccination or biosecurity measures.

Scaling up and intensification. Growing demand for livestock products has meant scaling up and intensification of livestock production and marketing systems. Intensive livestock farms and traditional, extensive holdings in proximity pose risks to each other since diseases emerge, spread and are controlled differently in each type of system. A disease agent may move from a dispersed population of wild animals or extensively kept livestock, into an intensive system, where the possibilities of spread are many times greater. In addition, if the newly susceptible animals are from a single genotype, the invading agent can move through the population quickly. It finds opportunities to transmit in order to promote its own survival, and continues to adapt in response to ineffective control strategies imposed by humans such as misuse of antibiotics.

A large intensive unit infected with a disease agent has the potential, if the disease escapes, to infect many other farms as disease is transmitted through the air, on vehicles and clothes and through market chains. Occasionally, a change to an existing intensively managed system creates the conditions for a disease agent to become more widespread in animals and pass to humans. When

the driver of disease is the production and marketing system rather than the natural environment or climate change, prevention requires proactive changes to livestock systems.

Habitat change. The interface between wildlife, humans and livestock is changing, as humans encroach on wildlife habitat, or habitat becomes degraded forcing wild animals to range further in search of food and water, or wildlife are used as food. As the contact between humans and wildlife becomes closer, it provides the opportunity for viruses such as SARS and avian influenza or influenza to jump species and, in some cases, become a new strain, gaining or decreasing in virulence as they spread within the new host niche. Direct impacts manifest as human sickness and death, but there also can be enormous indirect effects from efforts taken to contain the diseases. For example, measures that prevent the movement of animals, people or goods are hugely disruptive to global food chains and, in extreme cases, can have a short but significant impact on business, incomes and GDP. Health threats of this kind require excellent disease intelligence, timely reporting and the ability to mount a very rapid response should an outbreak begin.

products. Good business strategy is the key to survival. Urban populations are very vulnerable to instability in market chains.

The approach that China has taken to making its mega-cities reasonably food self-sufficient through zoning and subsidies may be one way to reduce vulnerability. Another is to limit monopolies and reliance on a few concentrated supply chains and, instead, to spread the sources of food so that many nations and regions supply many others.

Increased ethical concerns such as mitigating environmental damage and animal welfare requirements are beginning to affect livestock food supply. Currently the greatest effort in both of these areas is being undertaken in developed countries, particularly the EU (EUROPA, undated).

On the environment side, Brazil has invested in poultry production units with neutral impact on carbon emissions. It also recently moved to ban production of sugarcane in the Amazon area (BBC, 2009) and large supermarket chains and cattle companies have agreed to stop sourcing cattle from illegally cleared land (*Meat Trade*, 2009). China and some European countries have invested in biogas plants, as discussed earlier.

On the welfare side, the World Organisation for Animal Health (OIE) has introduced seven animal welfare standards for terrestrial animals covering transport, slaughter and culling (OIE, undated) and has a working group on animal welfare. Developing country governments have made limited investment in animal welfare initiatives, but there have been a number of special interest group initiatives.

If "green" initiatives gain wider traction, they will introduce new requirements into intensive production that may make it more expensive in the short term but should improve long-term sustainability.

BUILDING SUSTAINABLE SYSTEMS

If the changes outlined above were unidirectional and reasonably predictable, then it would be possible to adjust through changes in tech-

nology and management systems. But this is unlikely. We can expect climate events to become more frequent and more severe, with all of the related effects on health and markets. This variability hits smallholders and livestock dependent communities worse than intensive producers because their resources are already stretched, which limits their potential to withstand prolonged crises or adapt to new situations. Shoring up fragile societies indefinitely with emergency aid is not an option nor is leaving them to starve. Those who continue to live in marginal areas will need support in planning for their own future livestock production and sustaining their families and local communities. For growing urban populations, larger and more intensive systems better adapted to shocks are likely to be the main source of livestock protein for the future.

Animal health strategies carry useful lessons for food security. They do not assume that it is possible to predict and prepare for every change in conditions. However, well organized animal health systems have plans and resources in place to respond to surprises. An important consideration for food security is building in a sufficient margin for error. If a system is set up to use 100 percent of the available resources and produce at a high level in a "normal" year when things go well, then in a shock year, it will be badly hit and there will be a big drop in production. If this happens only once, the system will adjust, but if shocks happen often, there will be no reserve to draw on and eventually the system will be unable to recover. We can see this, for example, if rangelands are too heavily stocked to accommodate droughts and snowstorms and, at the same time, there is no built-in destocking process to allow the pasture to recover. The same is true for smallholder systems where the loss of crops or animals over several seasons leaves families with no safety net on which to draw.

Building preparedness into food systems requires changing the approach to risk analysis. This means planning production with wider margins of error and greater attention to what might happen if things fail, or emphasizing sustainable rather than short-term productivity to accommodate the possibility of failures or reduced levels of production over more than one production cycle. Rather than attempting to bring all to the highest level of productivity, a sustainable goal for mixed farming in particular might call for bringing lower performers up towards the middle. Some "slack" is needed in food systems to maintain stable food supplies in spite of extreme weather events and other supply disruptions. There may be benefits from intensification with limited concentration of production units, in order to reduce disease risk and environmental pollution, although this may be unpopular because of the infrastructure costs involved.

©FAO/Pius Utomi Ekpei

Conclusions

Livestock are important to the food security of millions of people today and, as shown in this review, will be important to the food security of millions more in the coming decades. Livestock source food is not essential to human nutrition but it is highly beneficial. In livestock systems that primarily consume roughage and agro-industrial waste products, livestock add to the food supply beyond what can be provided by crops. Moreover, they make a very important contribution to food access and stability through the income and products they provide to small-scale mixed farmers and pastoralists, the asset value of animals and their flexibility of use. The role that livestock play in feeding the future will be shaped by three distinct human populations, each with its own particular needs, namely: urban dwellers, small-scale mixed farmers and livestock-dependent populations.

URBAN CONSUMERS

The largest and fastest growing population lives in towns and cities, and its demand for reasonably priced meat, milk and eggs has been a strong inducement to intensify livestock food systems so that economies of scale can be realized and market chains managed efficiently. If current projections prove accurate, the largest growth in human population will remain in large urban centres, and the city populations will have even greater influence on the nature of demand for livestock products – the amount and type of livestock source food that is consumed, the way that farms and rangelands are managed, the distance products travel and the prices that farmers are paid.

Through its purchasing habits, this population has steadfastly supported global value chains for livestock and livestock products and, in turn, has benefited from intensive livestock production systems. Yet these are the same systems that currently cause great concern because of their emissions of greenhouse gases, pollution of water systems and competition for cereals. At the same time, small pockets of the urban population have driven "green" consumerism for livestock products through strongly voiced animal welfare and environmental concerns. Yet, as it stands, there are no technically or economically viable alternatives to intensive production for providing the bulk of the livestock food supply for growing cities. The future challenge is to factor environmental protection and system resilience into intensive livestock production.

Environmental challenge. An urgent challenge is to make intensive production more environmentally benign. Based on existing knowledge and technology, there are three ways to do this: reduce the level of pollution generated from greenhouse gases and manure; reduce the input of water and grain needed for each output of livestock protein; and recycle agro-industrial waste through livestock populations. All of these require capital investment and a supporting policy and regulatory environment.

Resilience challenge. Meeting the challenge of planning for food system resilience in a population that cannot feed itself requires a solid and stable production base for livestock source food. Higher food prices have encouraged investment in food production. This is potentially beneficial for urban food supply since it provides some scope to adapt and change, one of the conditions for resilience. Livestock diseases also must be dealt with, since intensive systems, and those that encroach upon forest environments or peri-urban areas without proper hygiene, are a fertile ground for new diseases, and many of them are managed in ways that are detrimental to animal health and welfare. It is not enough to pour funding into coping with the urgent disease threats of today – disease intelligence and epidemiological research must be financed to anticipate future diseases in the countries that produce the bulk of livestock source food.

Robust international trade systems also are essential to the resilience of food systems. City populations depend on trade for their food supply, and the production base can be hundreds of miles away. Governments have a vital role in securing and stabilizing trade agreements and promoting a sufficiently wide network of sources to act as a buffer against natural disasters and other shocks. Even where the foodshed for live-

stock products is wrapped closely around the urban population, as with Chinese megacities, the feed supplying the animal may be imported. There has been discussion recently (Von Braun and Torero, 2009) about the advisability of replenishing or re-establishing buffer stocks for food staples. Given the periodic instabilities in world supplies, this may be helpful. However, it is equally important for governments to look beyond their immediate national food self-sufficiency needs to the stability of the world supply.

PRODUCER–CONSUMERS

Mixed farmers and livestock-dependent populations, as producer-consumers, have different concerns from city populations. As suppliers of food to their own communities and contributors to the world food supply, they should benefit from investment in food systems and elevated prices. As excellent users of roughage and recyclers of waste, they make an important contribution to the food supply. However, they have a very limited ability to compete with large-scale intensive production.

Within small-scale and extensive systems, livestock make an important contribution to preserving food security, but people depending on these systems have very limited prospects to increase their income or expand their assets. This is evident from a rich-poor division that can be seen, for example, in the Horn of Africa, where some pastoralists have been forced to become contract herders because of economic circumstances (Aklilu and Catley, 2009) and in Mongolia where some herders with non-viable herd sizes have moved into cities.

Once this gap forms, it is extremely difficult to bridge. It is also evident from the numbers of small-scale producers who leave livestock production when competition pushes them out or when more secure off-farm opportunities beckon.

From a food security perspective, much of what can be said is already well known. Perhaps the most important argument to be made here is to stress the importance of rigorously applying a twin-track approach – dealing with short-term and long-term food insecurity issues in parallel.

Short-term response. The guiding principle for dealing with short-term shocks is to focus on protection of livestock assets. Households and communities able to maintain their assets during a crisis will be able to rebuild more easily when the shock is over. This might involve providing feed as well as food aid during a natural disaster, having a food security contingency plan as well as a disease-control contingency plan for dealing with major disease outbreaks, or using targeted culling during a disease outbreak to minimize asset destruction and the erosion of stocks of indigenous animals.

Long-term resilience. Dealing with long-term resilience for livestock-dependent populations and mixed farmers is a more difficult prospect than dealing with short-term shocks. These people undoubtedly benefit from the capital provided by their livestock. To grow economically, however, they need an institutional, policy and research environment that proactively supports them – as demonstrated by comparing the growth of cooperatives of small-scale producers in the Indian dairy subsector with the scaling-up of dairying in Brazil. Support in establishing access to the markets that offer longer term viability for smallholders, developing technology focussed on efficient use of roughage and by-products, and supporting land tenure and credit, particularly for women, can all help increase production from these systems and, thus, food access for those involved. Policies to promote the use of livestock for other economically valuable tasks, such as environmental services, also can improve the food security of their owners. However, in the end, there are no "magic bullets", and people may benefit most when livestock production is supported with parallel support in developing other livelihoods opportunities.

There are, therefore, two challenges with respect to livestock-dependent and small-scale mixed farmers. One is to make objective assessments of their contributions, based on social, economic and environmental factors, and to offer proactive support in activities, locations and economies where their contribution is greatest. There are examples of good practice from the field on which to build, although many of them are on a small scale. The other challenge is to manage the transition of those for whom livestock production is not a viable long-term prospect, by offering support and training to move into other livelihoods with more growth potential. However, this is a complex task, accompanied by considerable danger that the most vulnerable people will fall through the cracks, especially given the division of labour in most governments, research organizations and the international community.

A REGIONAL PERSPECTIVE

In all of the above, emerging economies will continue to play an important part as they have increasingly done for the past 40 years. Fan and Brzeska (2010) highlight the important role of emerging economies in global food security, which will depend not only on their capacity to produce but also their ability to invest wisely in their own rural societies, in agricultural research, rural infrastructure, markets and safety nets. The more advanced Latin American economies together with China, India and Russia have the potential to contribute a large percentage to both demand growth and future supply. These countries have all of the major production systems operational within their borders and are experiencing all of the food security challenges described in this report. They have a considerable capacity to produce food and potentially to stabilize supply, and a great deal of experience on which to draw in improving food access.

All of these countries are connected to global trade to varying degrees. They also are all urbanizing rapidly and will need to deal with an increasing challenge of feeding cities, which they currently handle in very different ways. All except India have some land into which to expand, although they also are looking for investment opportunities in other countries. All have the potential to make renewable energy from solar power or biofuels. All have growing economies that can provide public and private investment capital.

Latin America and China are moving in the direction of upscaling and intensification, meaning that the problems with intensive systems that have been described in this report will have to be solved in these countries. Russia is investing in intensive production and, as a relatively new investor, it has the opportunity to do this sustainably. India, with its high demand for dairy products and excellent local distribution networks, may be the place where innovation in small-scale mixed systems is taken furthest.

Africa barely participated in the livestock revolution yet now, in spite of widespread poverty and hunger, it is experiencing rapid demand growth for livestock source food, much of which has to be imported. A split is also developing in its livestock sector between the traditional production base which consists mainly of rangelands and small-scale farming, and a growing intensive poultry subsector near the cities. A number of constraints limit production levels and competitiveness of the livestock sector, including variable quality of feed supplies, water scarcity, food safety and inefficient trade within the continent that hampers its ability to pursue comparative advantage on a regional scale. However, with sufficient political will and some investment, there may be potential for African livestock production to make a larger contribution to food security in the continent than in the past.

WHO DOES WHAT?

Looking toward the future, it is obvious involvement in assuring the contribution of livestock to food security should come from across the board. The private and public sectors, food producers and consumers, research and technology development will all need to play a part.

Finance. Much of the growth in supply of livestock source foods will come from large-scale intensive systems in which the private sector is the main driver. The costs of changes to management to reduce environmental impact, improve efficiency and meet welfare standards are likely to be borne by the private sector for the most part, with some costs passed on to consumers in the price of food. Public sector finance is needed for basic infrastructure, and for research that takes a long-term view or that benefits the poor. It also can support animal health services in remote areas by contracting private providers to carry out government programmes. Public sector finance, both national and international, is also necessary to provide a temporary buffer to short, severe shocks during food crises.

Private sector foundations and NGOs that use both public and private finance can invest in initiatives that underpin the access of livestock dependent societies and small-scale mixed farmers to essential services. As systems change and some livestock keepers diversify or leave the sector altogether when they are unable to provide the quantity or quality demanded by the market, a combination of private and public finance will be needed to support them in developing specialized livestock enterprises, applying more efficient water management, carrying out pilot activities in environmental services or establishing new livelihoods outside of farming.

Policy, regulations and standards. Public regulation can enable the private sector to bring its efficiency and innovation into finding ways to improve the efficiency of livestock systems and their roles in recycling waste. We now are well aware, thanks to public sector overviews, that livestock are polluters. But we also have seen that the innovative private systems with the potential to feed the cities are capable of rising to the challenge of controlling pollution within intensive systems. In providing the policy to support the private sector and intensification, it is also critical to make sure the smallholder and extensive producers are not pushed aside. Policy also underpins the land-use patterns that influence the choices livestock keepers can make in rangeland management.

Public regulation and standards for animal health are strongly guided by the international animal health systems and embedded in international trade regulation through the sanitary and phytosanitary agreement of the World Trade Organization. In the same context, Codex Alimentarius, an international commission created by FAO and WHO defines global standards for food safety. However, in other aspects of livestock development important to the sustainability of food systems, such as environmental regulation, public regulation and standards are less well defined. In addition, they do not form part of international trade agreements, making their implementation more a matter for individual countries or companies and, in the future, for negotiation between public and private sectors. Policies guiding or supporting use of marginal lands and recycling the waste of other systems into on-the-hoof protein will also require negotiation between governments, private sector, civil society and local communities.

Research and technology. Some of the proposals and opportunities mentioned in this report will need research into technologies and institutions to increase understanding and generate knowledge that can guide the sector as well as national policy development. For example, improving the efficiency of livestock production may require developing breeds better adapted to particular production niches, while dealing with climate change and water stress will require finding ways to manage water more efficiently. Reducing environmental damage, developing innovative animal health systems and recycling waste all need new knowledge as well as ways of better applying existing knowledge.

Consumer choice and communication. Consumer choice will influence directions for livestock systems in terms of the products chosen and the way that animals are managed. Con-

sumers themselves are influenced by many forces, most of all their immediate social and peer groups. This means that public sector influence on good nutrition choices is limited, whether this means providing a balanced diet for children or not over-consuming livestock products. Governments can influence choices to some extent, through regulating what is provided in school meals and how foods are advertised, or through providing nutrition education. However, the rise in obesity over the past two decades would suggest that this has not been sufficiently effective. A more innovative and diverse approach to communicating about nutrition is obviously required, based on sound knowledge and relayed by respected individuals, peer groups and media.

Livestock's role in food security will not be driven by any one part of the livestock sector. It will depend on finding a way to create a coalition of all parties who in reality have very different backgrounds, responsibilities and goals but understand the big picture of what livestock has to offer to the world's food security and also of what it has to lose if they do not act together to ensure the sector has the tools it needs to sustain production at levels which meet the world's constantly increasing and changing demand.

References

ABARE. 2010. Australian Bureau of Agricultural and Resource Economics. Commodity statistics accessed from ABARE Web site (available at http://www.abare.gov.au/publications_html/data/data/data.html).

ACI. 2006. Agrifood Consulting. *Poultry Sector Rehabilitation Project – Phase I. The Impact of Avian Influenza on Poultry Sector Restructuring and its Socio-economic Effects.* Prepared for FAO by Agrifood Consulting International April 2006. Rome, FAO.

Agarwal, B. 1992a. The Gender and Environment Debate: Lessons from India. *Feminist Studies* 18(1):119-158.

Agarwal, B. 1992b. Gender Relations and Food Security. *In Unequal Burden: Economic Crises, Persistent Poverty, and Women's Roles* L. Beneria & S. Feldman, eds., Oxford, Westview Press.

Ahmed, A.U., Hill, R.V., Smith, L.C., Wiesmann, D.M. & Frankenberger, T. 2007. The world's most deprived: characteristics and causes of extreme poverty and hunger, 2020 Vision for Food, Agriculture and the Environment Discussion Paper No 43, International Food Policy Research Institute (IFPRI), Washington DC.

Ahuja, V., Gustafson, D., Otte, J. & Pica-Ciamarra, U. 2009. *Supporting Livestock Sector Development for Poverty Reduction.* PPLPI Research Report. Rome, Pro-Poor Livestock Policy Initiative, FAO.

Ahuja, V., Dhawan, M., Punjabi, M. & Maarse, L. 2008. *Poultry based livelihoods of the rural poor: case of Kuroiler in West Bengal.* Study Report. Doc 012. New Delhi, South Asia Pro-Poor Livestock Policy Programme, FAO and NDDB.

Aklilu, Y. & Catley, A. 2009. *Livestock Exports from the Horn of Africa: An Analysis of Benefits by Pastoralist Wealth Group and Policy Implications. R*eport commissioned by the Food and Agriculture Organization (FAO) under the Livestock Policy Initiative of the Intergovernmental Authority on Development (IGAD).

Alderman, H. & Behrman, J.R. 2003. *Estimated Economic Benefits of Reducing LBW in Low-Income Countries*, University of Pennsylvania: Philadelphia, PA.

Animal Husbandry Statistics. Ministry of Agriculture, Government of India.

Argwings-Kodhek, G., M'mboyi, F., Muyanga, M. & Gamba, P. 2005. *Consumption patterns of dairy products in Kenya's urban centres.* Conference paper. Tegemeo Institute of Agricultural Policy and Development, Egerton University, Nairobi, Kenya. Cited by Kaitibie *et al.,* (2008).

Arpi, E. 2006. India's Amul Dairy Cooperative.

Ashley, S. & Sandford, J. 2008. *Livestock Livelihoods and Institutions in the Horn of Africa.* IGAD LPI, Working Paper no. 10-08. Rome, FAO.

Asian Development Bank. 2010. *Gender Equality Results: Case Studies: Nepal.* Mandaluyong City, Philippines, Asian Development Bank, 2010.

Associated Press. 2008. Urban areas struggle to get grocers, fresh food (available at http://www.msnbc.msn.com/id/28300393/).

Bailey, D.V., Barret, C.B., Little, P.D. & Chabari, F. 1999. *Livestock Markets and Risk management among East African Pastoralists: a Review and Research Agenda.* GL-CRSP Pastoral Risk Management Project (PRMP), Technical Report 03/99.

Balakrishnan, V. 2002. *Developments in the Indian feed and poultry industry and formulation of rations based on local resources in in Protein Sources For The Animal Feed Industry.* FAO Expert Consultation and Workshop Bangkok, 29 April - 3 May 2002 (available at http://www.fao.org/docrep/007/y5019e/y5019e00.htm#Contents).

Bamire, A.S. & Amujoyegbe, B.J. 2004. Economics of Poultry Manure Utilization in Land Quality Improvement Among Integrated Poultry-Maize-Farmers in Southwestern Nigeria. *Journal of Sustainable Agriculture,* 1540-7578, 23(3) 2004, 21-37 Nigeria.

Barrow, E., Davies, J., Berhe, S. Matinu, V., Mohammed, N., Olenasha, W. & Rugadya, M. 2007. *Pastoral institutions for managing natural resources and landscapes.* IUCN Eastern Africa Office Policy Brief No. 3 (of 5), Nairobi. (available at http://cmsdata.iucn.org/downloads/pastoralist_institutions_for_managing_natural_resources_and_landscapes.pdf).

BBC. 2009. Brazil eyes Amazon sugar cane ban (available at http://news.bbc.co.uk/2/hi/8262381.stm).

Beef magazine. August 2008 (available at http://beefmagazine.com/markets/feed/0801-locking-your-corn/).

Bender, A. 1992. Meat and Meat Products in Human Nutrition in the Developing World. Food and Nutrition paper No. 53. Rome, FAO.

Bennett, A., Lhoste, F., Crook, J. & Phelan, J. 2006. The future of small scale dairying. In *The Livestock Report 2006* (A. McLeod, ed.). Rome, FAO.

Bentley, G.R., Aunger, R., Harrigan, A.M., Jenike, M., Bailey, R.C. & Ellison, P.T. 1999. Women's Strategies to Alleviate Nutritional Stress in a Rural African Society. *Social Science and Medicine*, 482:149-162.

BFREPA. 2010. British Free Range Egg Producers Association. *News headline: warning of further feed price increases* (available at http://www.theranger.co.uk/news/Warning-of-further-feed-price-increases_21517.html).

Birol, E., Roy, D. & Torero, M. 2010. *How Safe Is My Food? Assessing the Effect of Information and Credible Certification on Consumer Demand for Food Safety in Developing Countries.* IFPRI Discussion Paper 01029, October 2010, Washington D.C.

Black, K.D. 2001. *Environmental Impacts of Aquaculture.* Sheffield, Sheffield Academic, 2001.

Black, P.F, Murray, J.G. & Nunn, M.J. 2008. Managing animal disease risk in Australia: the impact of climate change. *Rev. sci. tech. Off. int. Epiz.*, 2008, 27(2), 563-580

Blackmore, E. & Keeley, J. 2009. *Understanding the social impacts of large-scale animal protein production.* Oxfam Novib / IIED preliminary scoping report as input to the Conference on the Social Impacts of the Large-Scale Meat and Dairy Production and Consumption. Oxfam-Novib.

Blench, R., Chapman, R. & Slaymaker, T. 2003. *A Study of the Role of Livestock in Poverty Reduction Strategy Papers (PRSPs).* PPLPI Working Paper no. 1. Rome, Pro-Poor Livestock Policy Initiative, FAO.

Blobaum, R. 1980. Biogas production in China. P. 212-216 in Biogas and Alcohol Fuels Production. Proceedings of a Seminar on Biomass Energy for City, Farm, and Industry. Emmaus, PA, The JG Press.

Bourdieu, P. 1977. *Outline of a Theory of Practice.* Cambridge, Cambridge University Press.

Brinkley, C. 2010. *Feeding Mega-Cities: Peri-urban Animal Agriculture.* Background paper prepared for the FAO Animal Production and Health Division. Unpublished.

Bumb, B. & Baanante, C. 1996, *World Trends in Fertilizer Use and Projections to 2020*, 2020 Brief No. 38. International Food Policy Research Institute, Washington, D.C., 1996.

Byarugaba, D.K., Nakavuma, J.L., Vaarst, M. & Laker, C. 2008. Mastitis occurrence and constraints to mastitis control in smallholder dairy farming systems in Uganda. *Livestock Research for Rural Development.* 20(1) (available at http://www.lrrd.org/lrrd20/1/byar20005.htm).

Capitalization of Livestock Program Experiences in India (CALPI). Undated. *Milk marketing in India: a review paper on the role and performance of informal sector.* CALPI (available at http://www.intercooperation.org.in/km/pdf/Documentation/Traditional/Informal%20milkmarket%20Desk%20study%20(Amit).pdf).

CARE. 1998. El Niño in 1997-1998: Impacts and CARE's Response.

CDC. 2010. *Operationalizing "One Health": A Policy Perspective – Taking Stock and Shaping an Implementation Roadmap.* (available at http://www.cdc.gov/onehealth/meetings.html).

Chacko, C.T., Gopikrishna, Padmakumar, V., Sheilendra, T. & Ramesh, V. 2010. India: growth, efficiency gains and social concerns. *In* P. Gerber, H. Mooney, J. Dijkman, S. Tarawali & C. de Haan, eds. *Livestock in a changing landscape, Vol. 2: Experiences and regional perspectives.* Washington, DC, Island Press.

Cohen, M.J. & Garrett, J.L. 2010. The food price crisis and urban food (in)security. *Environment and Urbanization* 222: 467-482. DOI: 10.1177/0956247810380375 www.sage-publications.com.

Committee on World Food Security. 2005. Thirty-first Session, Rome, 23-26 May 2005, Assessment of the World Food Security Situation, CFS: 2005/2.

Conant, R. 2010. Challenges and opportunities for carbon sequestration in grassland systems. A technical report on grassland management and climate change mitigation. *Integrated Crop Management* Vol. 9–2010. Rome, FAO.

Conant, R.T. & Paustian, K. 2002. Potential soil carbon sequestration in overgrazed grassland ecosystems. *Global Biogeochemical Cycles* 2002; 1143.

Costales, A. 2007. *Pig Systems, Livelihoods and Poverty: Current Status, Emerging Issues and Ways Forward.* PPLPI Research Report. Rome, Pro-Poor Livestock Policy Initiative, FAO.

Costales, A., Gerber, P. & Steinfeld, H. 2006. Underneath the Livestock Revolution. In *The Livestock Report 2006* (A. McLeod ed.). Rome, FAO. (available at www.fao.org/docrep/099/a0255e05.htm)

Costales, A., Otte, J. & Upton, M. 2005. *Smallholder Livestock Keepers in the Era of Globalization.* PPLPI Research Paper. Rome, Pro-Poor Livestock Policy Initiative, FAO.

De Weijer, F. undated, probably 2007. *Cashmere Value Chain Anlaysis Afghanistan.* Report prepared for the USAID Accelerating Sustainable Agriculture Programme.(available at http://www.ahdp.net/reports/Cashmere%20Value%20Chain%20Analysis.pdf).

DeBenoist, B., McLean, E., Egli, I. & Cogswell, M. 2008. *Worldwide prevalence of anaemia 1993–2005*: WHO global database on anaemia (available at http://whqlibdoc.who.int/publications/2008/9789241596657_eng.pdf).

Debrah, S.K. 1993. *Experiences in peri-urban dairy production, marketing and consumption. Cattle Research Network peri-urban dairy production project: pre-survey seminar.* Bamako, Mali 27-30. September, 1993. Cited by Cited by Smith, O.B. & Olaloku, E.A. (1998). *Peri-Urban Livestock Production Systems.* CFP Report 24 (available at http://idrc.ca/fr/ev-2513-201-1-DO_TOPIC.html.).

Delgado, C., Rosegrant, M., Steinfeld, H., Ehui, S. & Courbois, C. 1999. Livestock to 2020. The next food revolution. Food, Agriculture and the Environment discussion paper No 28, Washington DC, International Food Policy Research Institute, Rome, FAO, and Nairobi, International Livestock Research Institute.

Delgado, C.L. 2003. Rising Consumption of Meat and Milk in Developing Countries has Created a New Food Revolution. J. Nutr. 133:3907S-3910S, November 2003 (available at http://jn.nutrition.org/cgi/content/full/133/11/3907S; http://www.lrrd.org/lrrd21/9/betr21143.htm).

Delgado, C.L., Narrod, C.A. & Tiongco, M.M. 2008. *Determinants and Implications of the Growing Scale of Livestock Farms in Four Fast-Growing Developing Countries.* Research Report 157. International Food Policy Research Institute, FAO, LEAD.

Deutsch, L., Kalkenmark, M., Gordon, L., Rockstrom, J. & Folke, K. 2010. Water-mediated ecological consequences of intensification and expansion of livestock production. *In* H. Steinfeld, H. Mooney, F. Schneider & L. Neville, eds. *Livestock in a changing landscape, Vol. 1: Drivers, consequences, and responses.* Washington, DC, Island Press.

Dolberg, F. 2003. *A Review of Household Poultry Production as a Tool in Poverty Reduction with a Focus on Bangladesh and India.* PPLPI Working Paper no. 6. Rome, Pro-Poor Livestock Policy Initiative, FAO.

Ear, S. 2005. *The Political Economy of Pro-Poor Livestock Policy in Cambodia.* PPLPI Working Paper no. 26. Rome, Pro-Poor Livestock Policy Initiative, FAO.

Environment Agency, UK. Undated. Turning waste into valuable materials. (available at http://www.environment-agency.gov.uk/aboutus/wfo/epow/126982.aspx).

EUROPA. Undated. EU legislation on the protection of animals (available at http://ec.europa.eu/food/animal/index_en.htm accessed October 2010.)

Fafchamps, M., Udry, C., Czukas, K. 1998, Drought and saving in West Africa: are livestock a buffer stock, *Journal of Development Economics*, Vol.55(2) 273–305.

Fairfield, T. 2004. The Politics of Livestock Sector Policy and the Rural Poor in Bolivia, PPLPI Working Paper no. 15. Rome, Pro-Poor Livestock Policy Initiative, FAO.

Fan, S. & Brzeska, J. 2010. The Role of Emerging Countries in Global Food Security (available at http://www.ifpri.org/sites/default/files/publications/bp015.pdf).

FAO, UNICEF & UNDP. 2007. FAO/UNICEF/UNDP Report. *Joint Food Security Assessment Mission to Mongolia.* Ulaanbaatar, Mongolia. (available at http://www.fao.org/docrep/010/j9883e/j9883e00.HTM accessed January 2011.)

FAO. 1997. *Street foods.* FAO Food and Nutrition Paper No 63. Rome.

FAO. 1998. FAO's activities in relation to the 1997/98 El Niño (available at http://www.fao.org/english/newsroom/highlights/1998/elnino-e.htm).

FAO. 1999. Issues in urban agriculture. January 1999. (available at http://www.fao.org/ag/magazine/9901sp2.htm).

FAO. 2001. *Food for the Cities. Food supply and distribution policies to reduce urban food insecurity A briefing guide for Mayors, City Executives and Urban Planners in Developing Countries and Countries in Transition.* Food into Cities Collection, DT/43-00E. Rome.

FAO. 2003. *Understanding the Indigenous Knowledge and Information Systems of Pastoralists in Eritrea.* Rome, FAO (available at http://www.fao.org/docrep/006/y4569e/y4569e04.htm).

FAO. 2004. *Incorporating Nutrition Considerations into Development Policies and Programmes.* Policy Brief. Rome.

FAO. 2005. *Benefits and Potential Risks of the Lactoperoxidase system of Raw Milk Preservation.* Report of an FAO/WHO technical meeting. FAO Headquarters, Rome, Italy, 28 November - 2 December, 2005

FAO. 2006a. *Food Security Policy Brief.* June (2.) ftp://ftp.fao.org/es/ESA/policybriefs/pb_02.pdf accessed January 2010.

FAO. 2006b. Impacts of animal disease outbreaks on livestock markets. Introductory Paper on Animal Disease Outbreaks prepared for 21st Session of the Inter-Governmental Group on Meat and Dairy Products. Rome, Italy, 14 November 2006 (available at http://www.fao.org/docs/eims/upload//234375/ah670e00.pdf).

FAO. 2006c. World Agriculture towards 2030/2050. Interim report. Global Perspective Studies Unit. Rome, June.

FAO. 2006d. *State of World Aquaculture.* ftp://ftp.fao.org/docrep/fao/009/a0874e/a0874e04.pdf

FAO. 2007. Food Outlook Global Market Analysis - Poultry Meat (available at http://www.thepoultrysite.com/articles/918/food-outlook-global-market-analysis-poultry-meat.

FAO. 2008a *State of Food Insecurity in the World 2008: High food prices and food security – threats and opportunities.* FAO, Rome ftp://ftp.fao.org/docrep/fao/011/i0291e/i0291e00.pdf.

FAO. 2008b. Climate change, water and food security. Technical background document from the expert consultation held on 26 to 28 February 2008. Rome, FAO (available at http://www.fao.org/nr/water/docs/HLC08-FAOWater-E.pdf).

FAO. 2008c *The State of World Fisheries and Aquaculture 2008.* FAO Fisheries and Aquaculture Department, Rome, 2009 ftp://ftp.fao.org/docrep/fao/011/i0250e/i0250e01.pdf

FAO. 2009a. *State of Food Insecurity in the World 2009.* FAO, Rome.

FAO. 2009b. *Livestock in the balance.* State of Food and Agriculture 2009. FAO Rome.

FAO. 2009c. *Mapping traditional poultry hatcheries in Egypt.* Prepared by M. Ali Abd-El-hakim, Olaf Thieme, Karin Schwabenbauer and Zahra S. Ahmed. AHBL - Promoting strategies for prevention and control of HPAI. Rome.

FAO. 2010a. *State of Food Insecurity in the World 2010: Addressing food insecurity in protracted crises*, FAO Rome (available at http://www.fao.org/docrep/013/i1683e/i1683e.pdf).

FAO. 2010b. *The State of World Fisheries and Aquaculture.* Rome, FAO. 2010.

FAO. Undated. *Edible forest insects.* (available at http://www.fao.org/forestry/65422/en/).

FAO/OIE/WHO/UNSIC/UNICEF/World Bank. 2008. Contributing to One World, One Health: A Strategic Framework for Reducing Risks of Infectious Diseases at the Animal–Human–Ecosystems Interface. 14 October 2008 Consultation Document. ftp://ftp.fao.org/docrep/fao/011/aj137e/aj137e00.pdf

FARM-Africa. 2007. *Working Paper No 9. The Goat Model: A proven approach to reducing poverty among smallholder farmers in Africa by developing profitable goat enterprises and sustainable support services.* London: FARM Africa. (available at http://www.farmafrica.org.uk/resources/WP9%20The%20Goat%20Model.pdf)

Fiala, N. 2008. Meeting the demand: an estimation of potential future greenhouse gas emissions from meat production. Ecological Economics 2008; 67: 412-419.

Foeken, D. & Mwangi, A.M. Undated. *Increasing Food Security Through Urban Farming In Nairobi* (available at http://www.ruaf.org/sites/default/files/Nairobi.PDF).

Foeken, D. 2006. *Legislation, policies and the practice of urban farming in Nakuru, Kenya.* Contradictions abound. RUAF. (available at http://www.ruaf.org/sites/default/files/uam16_article8.pdf).

Gan, L. & Juan Yu. 2008. Bioenergy transition in rural China: Policy options and co-benefits. *Energy Policy* 36, p. 531-540.

Garcia, O., Hemme, T. & Khan, A.R. 2004a. *A Review of Milk Production in Bangladesh with Particular Emphasis on Small-Scale Producers.* PPLPI Working Paper no.7. Rome, Pro-Poor Livestock Policy Initiative, FAO.

Garcia, O., Hemme, T. & Mahmood, K. 2003. *A Review of Milk Production in Pakistan with Particular Emphasis on Small-Scale Producers.* PPLPI Working Paper no. 3. Rome, Pro-Poor Livestock Policy Initiative, FAO.

Garcia, O., Hemme, T. & Saha, A. 2004b. *The Economies of Milk Production in Orissa, India, with Particular Emphasis on Small-Scale Producers.* PPLPI Working Paper no.16. Rome, Pro-Poor Livestock Policy Initiative, FAO.

Garcia, O., Hemme, T., Huong Tra, H. & Tat Nho, L. 2006. *The Economics of Milk Production in Hanoi, Vietnam, with Particular Emphasis on Small-scale Producers.* PPLPI Working Paper no. 33. Rome, Pro-Poor Livestock Policy Initiative, FAO.

Geerlings, E., Albrechtsen, L. & Rushton J. 2007. *Highly pathogenic avian influenza: A rapid assessment of the socio-economic impact on vulnerable households in Egypt.* Report based on FAO and WFP livelihoods impact study of HPAI and its control in the Egyptian governorates of Assiut, Fayoum, Minya and Sohag, in partnership with UNDP's sister organisations BEST and Catholic Relief Services and in collaboration with the Egyptian Demographers Association. Rome, FAO.

Gerber, P., Chilonda, P., Franceshini, G., & Menzi, H. 2005. Geographical determinants and environmental implications of livestock production intensification in Asia. *Bioresource Technology* 96(13), 263-276.

Girardet, H. 1999. Urban farming and sustainable cities. Paper presented at International Workshop on the Policy Agenda, Havana, Cuba, 11–15 October 1999.

Gittelsohn, J., Thapa, M. & Landman, L.T. 1997. Cultural Factors, Caloric Intake and Micronutrient Sufficiency in Rural Nepali Households *Social Science & Medicine, 44 (11)*: 1739-1749.

Gning, M. 2005. Navigating the Livestock Sector: The Political Economy of Livestock Policy in Burkina Faso: PPLPI Working Paper no. 28. Rome, Pro-Poor Livestock Policy Initiative, FAO.

Griffin, M. 2004. Issues in the development of school milk. Paper presented at *School Milk Workshop, FAO Intergovernmental Group on Meat and Dairy Products.* Winnipeg, Canada, 17-19 June 2004. (available at http://www.fao.org/es/esc/common/ecg/169/en/School_Milk_FAO_background.pdf).

Guo, X., Mroz, T.A., Popkin, B.M., & Zhai, F. Structural Change in the Impact of Income on Food Consumption in China, 1989-93. Working Paper 99-02, Department of Economics University of North Carolina, Chapel Hill. (available at http://www.unc.edu/depts/econ/papers/99-02.pdf).

Gura, S. 2008. *Industrial livestock production and its impact on smallholders in developing countries. Consultancy Report to the League for Pastrolaist Peoples and Endonegnous Livestock Development.* April 2008. Cited by Blackmore and Keeley 2009.

Gurung, K., Man Tulachan, P. & Gauchan, D. 2005. *Gender and Social Dynamics in Livestock Management: A Case Study from Three Ecological Zones in Nepal.* (available at http://www.research4development.info/PDF/Outputs/Livestock/ZC0286-Case-Study-Nepal.pdf).

Halderman, M. 2005. *The Political Economy of Pro-Poor Livestock Policy-making in Ethiopia.* PPLPI Working Paper no. 19. Rome, Pro-Poor Livestock Policy Initiative, FAO.

Hancock, J. 2006. Exploring Impacts of Avian Influenza on Food Security. Internal working paper prepared for the FAO ECTAD socio-economic working group, November 2006. FAO, Rome.

Harding, J. 2010. What we are about to receive. Jeremy Harding (available at http://www.lrb.co.uk/v32/n09/jeremy-harding/what-were-about-to-receive/print)

Harvey, J. 2010. Plenty of guilt and a very heavy footprint. *In Business and Food Sustainability.* Financial Times Special Report (available at http://media.ft.com/cms/62017148-0b29-11df-9109-00144feabdc0.pdf).

Heimlich, R.E. & Bernard, C. 1993. *Agricultural adaptation to urbanization: Farm types in the United States Metropolitan Area.* USDA, Economic Research Service.

Heimlich, R.E. 1988. Metropolitan Growth and High-Value Crop Production. *In Vegetables and Specialties Situation and Outlook Report,* 17-26. TVS-244. U.S. Dept. of Agriculture, Economic Research Service. Washington.

Heimlich, R.E. & Brooks, D.H. 1989. Metropolitan Growth and Agriculture: Fanning in the City's Shadow. *AER-619.* U.S. Dept. of Agriculture, Economic Research Service. Washington, DC.

Hell, K., Fandohan, P., Ranajit Bandyopad-hyay, Kiewnick, S., Sikora, R. & Cotty, P.J. 2008. Pre- and post-harvest management of aflatoxin in maize: an African perspective. *In* L.J. Leslie, R. Bandyopadhyay, A. & Visconti, eds. *Mycotoxins: detection methods, management, public health and agricultural trade.* pp. 219-229. UK, CABI.

Henderson B., Gerber P. & Opio C. (in press) Livestock and climate change, challenges and options. Paper submitted to......

Hillbruner, C. & Murphy, M. 2008. *Food Security And Livelihoods In The Small Urban Centers Of Mongolia. Findings from the Aimag Center Food Security Assessment.* Final Report. Mongolia: Mercy Corps.

Hinrichs, J. 2006. Economic aspects of restructuring the poultry sector in Viet Nam. In: Poultry Sector Restructuring in Vietnam – Evaluation Mission. A study conducted by the Food and Agriculture Organization of the United Nations for the World Bank. July 2006.

Hoffmann, I. & Mohammed, I. 2004. The role of nomadic camels for manuring farmer's fields in the Sokoto close settled zone, northwest Nigeria . *Nomadic peoples* 8(1).

Hofmann, N. 2006. *A geographical profile of livestock manure production in Canada, 2006.* Environment Accounts and Statistics Division, Statistics Canada. (available at http://www.statcan.gc.ca/pub/16-002-x/2008004/article/10751-eng.htm).

Honhold, N. 1995. *Livestock Population and Productivity and the Human Population of Mongolia, 1930 to 1994* (Ulaanbaatar: European Union (Project ALA/MNG/9209), Ministry of Food and Agriculture).

Honhold, N. 2010. Mongolia: the limits of the last place on earth. Draft paper provided by the author to FAO as a reference for this publication, October 2010.

Hooper, R., Calvert, J., Thompson, R.L., Deetlefs, M.E. & Burney, P. 2008. Urban/rural differences in diet and atopy in South Africa. *Allergy 2008*: 63: 425–431 (available at http://onlinelibrary.wiley.com/doi/10.1111/j.1398-9995.2008.01627.x/pdf).

Hubbell, H.V. 1925. Land Subdivision Restrictions. *Landscape Architecture.* Vol 16. Oct. pp. 53-4.

Ibrahim, M., Porro, R. & Mauricio, R.M. 2010. Brazil and Costa Rica: deforestation and livestock expansion in the Brazilian legal Amazon and Costa Rica: Drivers, environmental degradation and policies for sustainable land management *In* P. Gerber, H. Mooney, J. Dijkman, S. Tarawali & C. de Haan, eds. *Livestock in a changing landscape, Vol. 2: Experiences and regional perspectives.* Washington, DC, Island Press.

ICASEPS. 2008. Indonesian Center for Agro-socioeconomic and Policy Studies. Livelihood and gender impact of rapid changes to bio-security policy in the Jakarta area and lessons learned for future approaches in urban areas. Rome, ICASEPS in collaboration with Food and Agriculture Organization.

IFAD. Undated. International Fund for Agricultural Development. China biogas project turns waste into energy. Web site (available at http://operations.ifad.org/web/guest/country/voice/tags/china/biogas

Ifft, J., Otte, J., Roland-Holst, D. & Zilberman, D. 2009. *Poultry Certification for Pro-Poor HPAI Risk Reduction.* Mekong Team Working Paper No. 6, HPAI Pro Poor Risk Reduction Project January 2009.

Ifft, J., Roland-Holst, D. & Zilberman, D. 2007. *Impact of Quality Characteristics on Demand for Chicken in Viet Nam.* PPLPI Research Report 07-14. Rome, Pro-Poor Livestock Policy Initiative, FAO.

IFPRI. 2004. Research and Outreach: Food Systems Governance. Washington DC: IFPRI.

Imai, K. 2003. Is Livestock Important for Risk Behaviour and Activity Choice of Rural Households? Evidence from Kenya. *Journal of African Economies*, 12(2): 271-295.

IMF. Undated. International Monetary Fund. Primary Commodity Prices, monthly data (available at http://www.imf.org/external/np/res/commod/index.asp).

Jabbar, M., Pratt, N. & Staal, S.J. 2008. Dairy *Development for the Resource Poor (Part 1,2,3)*, PPLPI Working Paper no. 44. Rome, Pro-Poor Livestock Policy Initiative, FAO.

Jackson, C. & Palmer-Jones, R. 1999. Rethinking Gendered Poverty and Work. *Development and Change*, 30 (3): 557-584.

Jackson, H.L. & Mtengeti, E.J. 2005. Assessment of animal manure production, management and utilization in Southern Highlands of Tanzania. *Livestock Research for Rural Development.* 17(10) (available at http://www.lrrd.org/lrrd17/10/jack17110.htm).

Jianming, C. 2003. Periurban Agriculture Development in China. Urban Agriculture. 9: 40-42.

Jones, N. 2010. A Taste of Things to Come. Nature 468, 752-753 2010 (available at http://www.nature.com/news/2010/101207/full/468752a.html).

Junfeng, L. 2007. REEEP: Activities that Support AD Project Development Worldwide. Paper presented at the Methane to Markets Partnership Expo, Beijing, 30 October – 1 November. (available at http://www.methanetomarkets.org/expo/docs/postexpo/ag_junfeng.pdf).

Kaiser, A. Undated. *Biogas from Energy Crop and organic waste in Europe in Germany*. Powerpoint presentation (available at http://www.business-meets-research.lu/fileadmin/user_upload/downloads/BMR_IBBK_A.KAISER.pdf).

Kaitibie, S., Omore, A., Rich, K., Salasya, B., Hooton, N., Mwero, D. & Kristjanson, P. 2008. *Influence pathways and economic impacts of policy change in the Kenyan dairy sector*. ILRI Research Report 15. Nairobi, International Livestock Research Institute (ILRI).

Kamuanga, M.J.B., Somda, J., Sanon, Y. & Kagoné, H. 2008. *Livestock and regional market in the Sahel and West Africa Potentials and challenges. Paris:* SWAC-OECD/ECOWAS 2008.

Kangmin, L. & Ho, M-W. 2006. *Biogas China*. Institute of Science in Society Report.(available at http://www.i-sis.org.uk/BiogasChina.php).

Kaplan, J.D., Johanssen, R.C. & Peters, M. 2004. The manure hits the land: economic and environmental implications when land application of nutrients is constrained. *Amer. J. Ag. Econ.* 86(3) (August 2004) 688-800

Kawashima, T. 2002. The use of food waste as a protein source for animal feed - current status and technological development in Japan. *In Protein Sources for the Animal Feed Industry.* FAO Expert Consultation and Workshop Bangkok, 29 April - 3 May 2002 (available at http://www.fao.org/docrep/007/y5019e/y5019e00.htm#Contents).

Kazianga, H. & Udry, C. 2006. Consumption smoothing? Livestock, insurance and drought in rural Burkina Faso. *Journal of Development Economics* 79 413–446.

Kazybayeva, S., Otte, J. & Roland-Holst, D. 2006. Livestock Production and Household Income Patterns in Rural Senegal, PPLPI Research Report. Rome, Pro-Poor Livestock Policy Initiative, FAO.

Ke, B. & Han, Y. 2007. Poultry sector in China: structural changes during the past decade and future trends. In *Poultry for the 21st Century. Avian influenza and beyond.* Bangkok, November 2007. Rome, FAO.

Ke, B. 2010. China: The East West Dichotomy. *In* P. Gerber, H. Mooney, J. Dijkman, S. Tarawali & C. de Haan, eds. *Livestock in a changing landscape, Vol. 2: Experiences and regional perspectives.* Washington, DC, Island Press.

Kenya Ministry of Agriculture. 2008. *Agriculture, Livestock, Fisheries and Rural Development Sector Medium-Term Plan 2008-2012.* Republic of Kenya.

Kharel, P. Undated. National alliance against hunger in Nepal. (available at http://www.iaahp.net/fileadmin/templates/iaah/pdf/NAAH_Nepal_EN.pdf).

Knips, V. 2005. Developing Countries and the Global Dairy Sector: Part 1: Global Overview: PPLPI Working Paper no. 30. Rome, Pro-Poor Livestock Policy Initiative, FAO.

Knips, V. 2006 Developing Countries and the Global Dairy Sector: Part II: Country Case Studies: PPLPI Working Paper no. 31. Rome, Pro-Poor Livestock Policy Initiative, FAO.

Krishna, A., Kristjanson, P., Nindo, W. & Radeny, M. 2004. *Pathways out of Poverty in Western Kenya and the Role of Livestock.* PPLPI Working Paper no. 14. Rome, Pro-Poor Livestock Policy Initiative, FAO.

LEAD. Undated. Livestock Production Systems Classification: Mixed systems (available at http://www.fao.org/ag/againfo/programmes/en/lead/toolbox/Refer/ProSystR.htm#Mixed).

Lee-Smith, D. & Memon P.A. 1994. Urban agriculture in Kenya. *In* A.G.Egziabher *et al.,* eds. *Cities feeding people: an examination of urban agriculture in East Africa.* Ottawa: International Development Research Centre.

Lee-Smith, D. 2010. Cities feeding people: an update on urban agriculture in equatorial Africa. *Environment & Urbanization.* 22(2): 483–499.

LEGS. 2009. Livestock Emergency Guidelines and Standards. UK, Practical Action Publishing. (available at http://www.livestock-emergency.net/userfiles/file/legs.pdf).

Leonard, W.R. 1991. Household-level Strategies for Protecting Children from Seasonal Food Scarcity. *Social Sci Med.* 33(10): 1127-1133.

Leyland, T. & Akabwai, D.M.O. Undated. *Delivery of Private Veterinarian Supervised Community-Based Animal Health Services To Pastoralist Areas Of The Greater Horn Of Africa.* UK: Vetwork.

Lipton, M. & Longhurst, R. 1989. *New Seeds and Poor People.* London, Routledge.

Lopez, R.A., Adelaja, A.O. & Andrews, M.S. 1988. The Effects of Suburbanization on Agriculture. *American Journal of Agricultural Economics.* 70 (2), 346-358.

LPP, LIFE Network, IUCN–WISP and FAO. 2010. *Adding value to livestock diversity – Marketing to promote local breeds and improve livelihoods.* FAO Animal Production and Health Paper. No. 168. Rome.

Ly, C., Fall, A. & Okike, I. 2010. West Africa: the livestock sector in need of regional strategies. *In* P. Gerber, H. Mooney, J. Dijkman, S. Tarawali & C. de Haan, eds. *Livestock in a changing landscape, Vol. 2: Experiences and regional perspectives.* Washington, DC, Island Press.

Makkar, H. FAO Animal Production and Health Division, personal communication.

Maltsoglou, I. & Rapsomanikis, G. 2005. *Contribution of Livestock to Household Income: A Household Typology Based Analysis, Vietnam.* PPLPI Working Paper no. 21. Rome, Pro-Poor Livestock Policy Initiative, FAO.

Maltsoglou, I. & Taniguchi, K. 2004. *Poverty, Livestock and Household Typologies in Nepal.* PPLPI Working Paper no. 13. Rome, Pro-Poor Livestock Policy Initiative, FAO.

Maltsoglou, I. 2007. *Household Expenditure on Food of Animal Origin: A Comparison of Uganda, Vietnam and Peru.* PPLPI Working Paper no. 43. Rome, Pro-Poor Livestock Policy Initiative, FAO.

Maltsoglou, I. & Taniguchi, K. 2004, Poverty, Livestock and Household Typologies in Nepal, PPLPI Working Paper No.13 (available at http://www.fao.org/ag/againfo/programmes/en/pplpi/docarc/wp13.pdf)

Mamo, G. 2007. Community? Forest Management in Borana. *In* A. Ridgewell, G. Mamo and F. Flintan, eds. *Gender & Pastoralism Vol. 1: Rangeland & Resource Management in Ethiopia*. Addis Abbaba: SOS Sahel Ethiopia.

Martínez, J., Jaro, P.J., Aduriz, G., Gómez, E.A., Peris, B. & Corpa, J.M. 2007. Carcass condemnation causes of growth retarded pigs at slaughter. *The Veterinary Journal* 174(1), July pp 160-164.

Maxwell, D.G. 1994. The household logic of urban farming in Kampala. *In Cities Feeding People; an examination of urban agriculture in East Africa*. International Development Research Centre. Ottawa, Canada. p. 146. Cited by Smith, O.B. and Olaloku, E.A. 1998. *Peri-Urban Livestock Production Systems*. CFP Report 24. (available at http://idrc.ca/fr/ev-2513-201-1-DO_TOPIC.html.)

Maykuth, A. 1988. Pigging Out. *In Philadelphia Inquirer, September 20, 1988*. (available at http://www.maykuth.com/Archives/pigs88.htm)

Mcleod, A. & de Haan, N. 2009. Mission report to the Avian Influenza Team in Viet Nam. FAO internal document.

McLeod, A. 2009. The Economics of Avian Influenza. *In* D. E. Swayne, ed, *Avian Influenza*. Blackwell Publishing Ltd., Oxford.

McLeod, A., Honhold, N. & Steinfeld, H. 2010. Responses on emerging livestock diseases *In* H. Steinfeld, H. Mooney, F. Schneider & L. Neville, eds. *Livestock in a changing landscape, Vol. 1: Drivers, consequences, and responses*. Washington, DC, Island Press.

McLeod, A., Taylor, N., Lan, L.T.K., Thuy, N.T., Dung, D.H. & Minh, P.Q. 2002. *Control of classical swine fever in the Red River Delta of Vietnam: A stakeholder analysis and assessment of potential benefits, costs and risks of alternative proposals.* Report on Phase 1 of the study. Hanoi: EUSVSV Project.

McMichael, A.J., Powles, J.D., Butler, C.D. & Uauy, R. 2007. Food, livestock production, energy, climate change and health. Energy and Health 5. In *The Lancet* Published Online. September 13, 2007DOI:10.1016/S0140-6736(07)61256-2.

Meat and Livestock Australia. 2001. Web site accessed Jan 2011 (available at http://www.mla.com.au/Prices-and-markets/Trends-and-analysis/Sheep-and-goats/Domestic-consumption).

Meat Trade News Daily. 2009. Brazil - Carrefour and Wal-Mart ban Amazon related beef. 16 June 2009 (available at http://www.meattradenewsdaily.co.uk/news/160609/brazil___carrefour_and_wal_mart_ban_amazon_related_beef.aspx).

Mehta, R., Nambiar, R.G., Delgado, C.L. & Subramanyam, S. 2003. *Policy, Technical, and Environmental Determinants and Implications of the Scaling-Up of Broiler and Egg Production in India*. Final Report of IFPRI-FAO Livestock Industrialization Project: Phase II. Washington D.C., IFPRI and Rome, FAO.

Mhlanga, F.N., Khombe, C.T. & Maukza, S.M. 1999. *Indigenous livestock types of Zimbabwe*. Department Of Animal Science University Of Zimbabwe (available at http://www.ilri.org/InfoServ/Webpub/fulldocs/AnGenResCD/docs/IndiLiveGenoZimbabwe/TableofContents.htm#TopOfPage).

Micronutrient Initiative. 2009. Investing in the future: a united call to action on vitamin and mineral deficiencies. (available at http://www.unitedcalltoaction.org/documents/Investing_in_the_future.pdf).

Mission East. 2010. Project report to FAO on results of a cost-sharing campaign for brucellosis control in Tajikistan. Unpublished.

Morgan, N. & Tallard, G. Undated. *Cattle And Beef International Commodity Profile.* Background paper for the Competitive Commercial Agriculture in Sub–Saharan Africa (CCAA) Study (available at http://siteresources.worldbank.org/INTAFRICA/Resources/257994-1215457178567/Cattle_and_beef_profile.pdf).

Morgan, N. Undated. *Repercussions of BSE on International Meat Trade: Global Market Analysis* (available at http://www.fao.org/ag/aGa/agap/FRG/Feedsafety/pub/morgan%20bse.doc).

Mukhebi, A.W. 1996. Assessing economic impact of tick-borne diseases and their control: the case of theileriosis immunisation. In A.D. Irvin, J.J. McDermott & B.D. Perry, eds *Epidemiology of ticks and tick-borne diseases in Eastern, Central and Southern Africa.* Proc. Workshop, 12-13 March, Harare. Nairobi, International Livestock Research Institute (ILRI).

Nakiganda, A., Mcleod, A., Bua, A., Phipps, R., Upton, M. & Taylor, N. 2006. Farmers' constraints, objectives and achievements in smallholder dairy systems in Uganda. *Livestock Research for Rural Development.* 18(69) (available at http://www.lrrd.org/lrrd18/5/naki18069.htm).

NaRanong, V. 2007. Structural Changes in Thailand's Poultry Sector and its Social Implications. A publication commissioned by the FAO-AGAL, pp.37. In *Poultry In The 21st Century: Avian Influenza And Beyond.* International Poultry Conference. Bangkok, November 2007. ISBN 978-92-5-106063-6 Rome, FAO.

National Research Council. 2009. Emerging technologies to benefit farmers in sub-Saharan Africa and South Asia. Washington DC. National Academies Press.

National Statistical Office of Mongolia. 1980 to 2009. Annual yearbooks.

National Statistical Office of Mongolia. 2007. Statistical Yearbook.

NDDB. 2010. National dairy statistics presented on the Department of Animal Husbandry, Dairying & Fisheries Web site, update May 2010 (available at http://dahd.nic.in/).

Neuman, C.G. & Harris, D.M. 1999. Contribution of animal source foods in improving diet quality for children in the developing world. Washington DC: Prepared for World Bank.

Neumann, C., Demment, M.W., Maretzki, A., Drorbaugh, N. & Galvin, K. 2010. Chapter 12. The Livestock Revolution and Animal Source Food Consumption. *In* Steinfeld H., Mooney H., Schneider F., & Neville L., eds. *Livestock in a changing landscape, Vol. 1: Drivers, consequences, and responses.* Washington, DC, Island Press.

Ngutu, M.N., Lelo, F.K., Kosgey, I. & Kauffmann, B. Undated. *Collective community management of livestock drug trade in a pastoralist setting: a case study of Ramati and Maslis groups of the greater Marsabit district.* Report to the Kenya Agricultrual Research Intitute.

Nyungu, P. & Sithole, L. 1999. *Analysis of Dynamic Gender Relations around Goat Acquisition, Ownership and Disposal in Chizumba Communal Area, Mwenezi District, Zimbabwe.* Makoholi Experiment Station, Masvingo, Zimbabwe. Ministry of Agriculture.

OECD-FAO. 2010. *Agricultural Outlook 2010-2019* (available at http://www.agri-outlook.org/document/10/0,3746,en_36774715_36775671_42852746_1_1_1_1,00.html).

OIE. Undated. The OIE's objectives and achievements in animal welfare (available at http://www.oie.int/animal-welfare/key-themes/ accessed February 2011.)

Okali, C. 2009. Making women partners in the livestock revolution. Unpublished background paper for FAO's State of Food and Agriculture 2010, in press.

Okike, I., Spycher, B., Williams, T.O. & Baltenweck, I. 2004. Lowering cross-border livestock transportation and handling costs in West Africa. ILRI/CFC/FAO/CILSS—West Africa Livestock Marketing: Brief 3. Nairobi (Kenya): ILRI.

Omore, A., Muriuki, H.G.M., Kenyanjui, M., Owango, M. & Staal, S. 2004. *The Kenyan dairy subsector: a rapid appraisal.* SDP research and development report. SDP (Smallholder Dairy Project), Nairobi, Kenya (available at http://www.smallholderdairy.org).

Oonincx, D.G.A.B., van Itterbeeck, J., Heetkamp, M.J.W., van den Brand, H. & van Loon, J.J.A. 2010. An Exploration on Greenhouse Gas and Ammonia Production by Insect Species Suitable for Animal or Human Consumption. *PLoS ONE 5(12): e14445.* doi:10.1371/journal.pone.0014445.

Opio, C. 2007. Background paper for H. Steinfeld, H. Mooney, F. Schneider & L. Neville, eds. *Livestock in a changing landscape, Vol. 1: Drivers, consequences, and responses.* Washington, DC, Island Press.

Otte, J. 2006. The Hen Which Lays Golden Eggs: Why backyard poultry are so popular. PPLPI Features. July 2006. Rome, Pro-Poor Livestock Policy Initiative, FAO. (available at http://www.fao.org/ag/againfo/programmes/en/pplpi/docarc/feature01_backyardpoultry.pdf).

Otte, J., Pica-Ciamarra, U. & Roland-Holst, D. 2008. Food Markets and Poverty Alleviation, PPLPI Research Report. Rome, Pro-Poor Livestock Policy Initiative, FAO.

Over, H.J., Jansen, J. & van Olm, P.W. 1992. Distribution and impact of helminth diseases of livestock in developing countries. FAO Animal Production and Health Paper No. 96. FAO, Rome.

Owens, G.M. 2007. Analyzing impacts of bioenergy expansion in China using strategic environmental assessment. *Management of Environmental Quality* 18(4), p. 398-407.

PAHO (Pan American Health Organization). 2006. *Assessing the economic impact of obesity and associated chronic diseases: Latin America and the Caribbean.* Fact Sheet, April 2006. Washington, DC.

Parthasarathy, P.R. and Birthal, P.S. 2008. Typology of Mixed Crop-Livestock Systems in Nepal. In *Livestock in mixed farming systems in South Asia.* New Dehli, National Centre for Agricultural Economics and Policy Research, and India, International Crops Research Institute for the Semi-Arid Tropics.

Pavanello, S. 2010. *Livestock marketing in Kenya-Ethiopia border areas: A baseline study.* HPG Working Paper. July 2010.

Perry, B. & Sones K. 2008. Global livestock disease dynamics over the last quarter century: drivers, impacts and implications. Background paper prepared for the FAO (2009b).

Petrie, O.J. 1995 "End Uses of Wool" in *Harvesting Of Textile Animal Fibres.* FAO Agricultural Services Bulletin No. 122. Rome, FAO (available at http://www.fao.org/docrep/v9384e/v9384e04.htm).

Pica-Ciamarra, U. & Otte, J. 2009. *Poultry, Food Security and Poverty in India: looking beyond the farm gate.* PPLPI Research Report. Rome, Pro-Poor Livestock Policy Initiative, FAO.

Pica-Ciamarra, U., Tasciotti, L., Otte, J. & Zezza, A. (in preparation) *Livestock and Livelihoods in Developing Countries: Evidence from Household Surveys.* Draft paper.

Pingali, P., Alinovi, L. & Sutton, J. 2005. Food security in complex emergencies: enhancing food system resilience *In Disasters,* 29(s1) UK & USA, Blackwell Publishing.

Pinstrup-Andersen, P. 2009. Food security: definition and measurement. *In Food Sec (2009) 1:5-7.*

Pinstrup-Andersen, P., Burger, S., Habicht, J.P. & Peterson, K. 1993. Protein-Energy Malnutrition. *In Disease Control Priorities in Developing Countries*, eds. Jamison D.T., Mosley W.H., Measham A.R., and Bobadilla J.L., pp. 391–420. New York, Oxford University Press for the World Bank.

Popkin, B.M., Horton, S. & Kim S. 2001. *The nutritional transition and diet-related chronic diseases in Asia: Implications for prevention.* FCND Discussion Paper No. 105. Food Consumption and Nutrition Division, International Food Policy Research Institute, 2033 K Street, N.W., Washington, D.C. 20006.

Povcal. *(available at http://iresearch.world-bank.org/PovcalNet/povDuplic.html. (using poverty line of 1.25$ per day in PPP or 38 $ per month).*

Pozzi, F. & Robinson, T. 2007. Poverty and Welfare Measures in the Horn of Africa: IGAD LPI, Working Paper no. 08-08. Rome, FAO.

Public Health of Canada. 2009. One World One Health: from ideas to action. Report of the expert consultation. March 16-19, Winnipeg, Manitoba.

Quisumbing, A.R., Meinzen-Dick, R.S. & Smith, L.C. 2004. Increasing the effective participation of women in food and nutrition security in Africa. Africa Conference Brief 4. International Food Policy Research Institute. Washington.

Quisumbing, A.R., Brown, L.R., Feldstein, H.S., Haddad, L. & Pena, C. 1995. *Women: The Key to Food Security.* Food Policy Statement No. 21. Washington, DC, IFPRI.

Raas, N. 2006. *Policies and Strategies to Address the Vulnerability of Pastoralists in Sub-Saharan Africa.* PPLPI Working Paper No. 37. Rome, Pro-Poor Livestock Policy Initiative, FAO.

Rae, A. & Nayga, R. 2010. Chapter 2 Trends in consumption, production and trade in livestock and livestock products. *In* H. Steinfeld, H. Mooney, F. Schneider & L. Neville, eds. *Livestock in a changing landscape, Vol. 1: Drivers, consequences, and responses.* Washington, DC, Island Press.

Reardon, T., Timmer, C.P. & Minten, B. 2010. *Supermarket revolution in Asia and emerging development strategies to include small farmers.* PNAS Early edition (available at http://www.pnas.org/content/early/2010/12/01/1003160108.full.pdf).

Reardon, T., Timmer, C.P., Barrett, C.B. & Berdegue, J.A. 2003. The rise of supermarkets in Africa, Asia, and Latin America. *Am J Agric Econ* 85:1140–1146.

Regmi, A. & Dyck, J. Undated. Effects of Urbanization on Global Food Demand in *Changing Structure of Global Food Consumption and Trade/WRS-01-1* USA: Economic Research Service/USDA.

Rosegrant, M.W., Cai, X. & Cline, S.A 2002. *Global water outlook to 2025, averting an impending crisis.* A 2020 vision for food, agriculture, and the environment initiative. Washington, DC: IFPRI and IWMI.

Rosenzweig, M.R. & Wolpin, K.I. 1993. Credit market constraints, consumption smoothing, and the accumulation of durable production assets in low-income countries: investments in bullocks in India. *Journal of Political Economy 101 (2), 223–244.*

Ross, J. & Horton, S. 1998. *Economic consequences of iron deficiency.* Ottawa: The Micronutrient Initiative.

Ru-Chen, C. 1981. The Development of Biogas Utilization in China. *Natural Resources Forum 5(3),* p. 277-282.

Rymer, C. 2006. Increasing the contribution that goats make to the livelihoods of resource-poor livestock keepers in the Himalayan Forest Region. Final Technical Report to Natural Resources International Ltd, UK. Retrieved January 2011. (available at http://www.dfid.gov.uk/r4d/PDF/outputs/R7632d.pdf).

Salasya, B., Rich, K., Baltenweck, I., Kaitibie, S., Omore, A., Murithi, F., Freeman, A. & Staal, S. 2006. *Quantifying the economic impacts of a policy shift towards legalizing informal milk trade in Kenya.* ILRI (International Livestock Research Institute) Discussion Paper 1. ILRI, Nairobi, Kenya.

Schlebecker, T. 1967. *A History of American Dairying.* Chicago: Rand McNally. 1967.

Schneider Group. Undated. Market indicators for cashmere from 1998 to 2012 (available at http://www.gschneider.com/index.php?page=marketindicators)

Sere, C. & Steinfeld, S. 1996. World Livestock Production Systems: Current status, issues and trends (available at http://www.fao.org/ag/againfo/programmes/en/lead/toolbox/Paper127/cover1.htm).

Sharma, V.P. 2010. Animal traction in South Asia. Background paper prepared for FAO's Regional Office for Asia and the Pacific, unpublished.

Smith, O.B. & Olaloku, E.A. 1998. Peri-Urban Livestock Production Systems. CFP Report 24 (available at http://idrc.ca/fr/ev-2513-201-1-DO_TOPIC.html).

Sponberg, K. 1999. Weathering a storm of global statistics. *Nature, 400,* p. 13. cited by **UNEP/NCAR/UNU/WMO/ISDR.** 2000. Lessons Learned from the 1997–98 El Niño: Once Burned, Twice Shy? (available at http://www.ccb.ucar.edu/un/enFinal.pdf).

Stage, J., Stage, J. & McGranahan, G. 2010. Is urbanization contributing to higher food prices? *Environment & Urbanization* Vol 22(1): 199–215. DOI: 10.1177/0956247809359644.

Starkey, P. 2010. Livestock for traction: world trends, key issues and policy implications. AGA working paper series. Rome, FAO.

Steinfeld, H., Gerber, P., Wassenaar, T., Castel, V., Rosales, M., De Haan, C. 2006 *Livestock's Long Shadow: Environmental Issues and Options.* Rome, FAO.

Steinfeld, H., Gerber, P. & Opio, C. 2010. Responses on environmental issues. *In* H. Steinfeld, H. Mooney, F. Schneider & L. Neville, eds. *Livestock in a changing landscape, Vol. 1: Drivers, consequences, and responses.* Washington, DC, Island Press.

Stokstad, E. 2004. Salmon survey stokes debate about farmed fish. *Science 2004*; 303(5655): 154–5.

Stuart, T. 2009. *Waste: Uncovering the global food scandal.* London, Penguin Books.

Swiss Re. 1999. El Niño 1997/98: On the phenomenon's trail. Zurich, Switzerland: Swiss Reinsurance, 8 pp. (available at http://www.swissre.com/).

TECA. Undated. *Mastitis control in smallholder dairy cows.* (available at ://www.fao.org/teca/node/4656).

The Economist. 2009. Egypt's pigs: What a Waste. May 7, 2009. (available at http://www.economist.com/node/13611723?story_id=13611723).

The Economist. 2010. Slaughterhouse Rules. Jun 24th 2010. (available at http://www.economist.com/node/16436481).

The *In Vitro* Meat Consortium. 2008. Preliminary Economics Study. Project 29071 V5 March 2008. (available at http://invitromeat.org/images/Papers/invitro%20meat%20economics%20study%20v5%20%20march%2008.pdf)

Thompson, E., Harper, A.M. & Kraus, S. 2008. Think Globally, Eat Locally. San Fransisco Foodshed Assessment. American Farmland Trust.

Thornton, P. 2010. Livestock production: recent trends, future prospects. *Phil. Trans. R. Soc. B* 2010 365, 2853-2867.

Thornton, P.K. & Herrero, M. 2010. Potential for reduced methane and carbon dioxide emissions from livestock and pasture management in the tropics. *Proceedings of the National Academy of Sciences* USA, 107: 19627-19632

Thornton, P.K. & Gerber, P. 2010. Climate change and the growth of the livestock sector in developing countries *Mitigation and Adaptation Strategies for Global Change* (in press).

Thornton, P.K., Kruska, R.L., Henninger, N., Kristjanson, P.M., Reid, R.S., Atieno, F., Odero, A.N. & Ndegwa, T. 2002. *Mapping poverty and livestock in the developing world.* ILRI (International Livestock Research Institute), Nairobi, Kenya (available at http://www.ilri.org/InfoServ/Webpub/fulldocs/mappingPLDW/index.htm).

Thøy, K., Wenzel, H., Jensen, A.P. & Nielsen, P.

2009. Biogas from manure represents a huge potential for reduction in global greenhouse gas emissions. *2009 IOP Conf. Ser.: Earth Environ. Sci.* 6 242020.

Tiong, C.K. & Bing, C.S. 1989. Abattoir condemnation of pigs and its economic implications in Singapore British Veterinary Journal Volume 145(1), January-February, pp.77-84.

Tisdell, C.A., Harrison, S.R. & Ramsay, G.C. 1999. The economic impacts of endemic diseases and disease control programmes *Rev. sci. tech. Off. int. Epiz.*, 1999, 18(2), 380-398.

Tung, D.X. 2005. Smallholder Poultry Production in Vietnam. Marketing Characteristics and Strategies. *In: NSPD 2005* (available at http://www.poultry.kvl.dk/upload/poultry/workshops/w25/papers/tung.pdf).

Umar, A. & Baulch, B. 2007. *Risk Taking for a Living Trade and Marketing in the Somali Region of Ethiopia*, UN OCHA-PCI, April 2007.

UN HABITAT. 2010. *Urban Indigenous Peoples and Migration: A Review of Policies, Programmes and Practices.* United Nations Housing Rights Programme Report No. 8. Nairobi.

UN Standing Committee on Nutrition. 2005. Overweight and Obesity: A New Nutrition Emergency? SCN News No. 29.

UN Standing Committee on Nutrition 2010. 6th Report on the World Nutrition Situation (available at http://www.unscn.org/files/Publications/RWNS6/report/SCN_report.pdf).

UNDP. 2009 *Human Development Report 2009*. New York.

UNFPA. 2007. *State of the World Population 2007*. New York.

UNFPA. 2009. *State of the World Population 2009*. New York.

UNFPA. 2010. *State of World Population 2010*. New York.

UNICEF. 2007. *The State of the World's Children 2007*. New York.

UNICEF. Undated. *Vitamin and Mineral Deficiency, A Global Progress Report* (available at http://www.micronutrient.org/CMFiles/PubLib/VMd-GPR-English1K-WW-3242008-4681.pdf).

United Nations Population Division, 2009. *World Population Prospects: The 2008 Revision.* (available at http://esa.un.org/unpp).

Upton, M. 2004. *The Role of Livestock in Economic Development and Poverty Reduction.* PPLPI Working Paper no. 10. Rome, Pro-Poor Livestock Policy Initiative, FAO.

USAID. 2007. USAID Supports Marketing of Traditionally-raised Poultry to Fight Bird Flu. USAID Viet Nam press release Thursday, March 05, 2009.

USAID. 2009. USAID Executive Brief: El Niño and Food Security in Southern Africa October 2009. The Famine Early Warning Systems Network. (available at http://www.fews.net/docs/Publications/El_Nino_brief_South_Oct_2009_final.pdf).

Von Braun, J. & Torero, M. 2009. Implementing Physical and Virtual Food Reserves to Protect the Poor and Prevent Market Failure. IFPRI Policy Brief 10, February 2009. (available at http://www.ifpri.org/sites/default/files/publications/bp010.pdf).

Von Braun, J. 2008. *Rising Food Prices: What Should be Done?* IFPRI Policy Brief. Washington : International Food Policy Research Institute. (available at http://www.ifpri.org/publication/rising-food-prices).

Walker, A. 2010. UN calls meeting on food price concerns. BBC World Service. (available at http://www.bbc.co.uk/news/business-11177346).

Walker, J. 2006. Utilizing an Untapped Resource; Manure Use in The Bolivian Altiplano as a Means to Increase Overall Production. SANREM CRSP (available at http://pdf.usaid.gov/pdf_docs/PNADL268.pdf).

Walker, P., Rhubart-Berg, P., McKenzie, S., Kelling, K. & Lawrence, R.S. 2005. Public health implications of meat production and consumption. *Public Health Nutrition*: 8(4), 348–356.

Walmart. 2010. Walmart's Integrated Organic

Diversion Program 2010. *Solid Waste and Recycling Programs.* (available at http://www.epa.gov/epawaste/rcc/resources/meetings/rcc-2010/fanning.pdf).

WHO. 2000. *Manual on the Management of Nutrition in Major Emergencies.* IFRC/UN-HCR/WFP/WHO. Geneva: WHO.

WHO. 2001. *Water, Sanitation and Health: Water Related Diseases* (available at http://www.who.int/water_sanitation_health/diseases/malnutrition/en.) Accessed January 2010.

WHO. 2003. Diet, nutrition and the prevention of chronic diseases: report of a joint WHO/FAO expert consultation, Geneva, 28 January – 1 February 2002

WHO. 2007. The challenge of obesity in the WHO European region and the strategies for response. F. Branca, H. Nikogosian and T. Lobstein eds. Geneva: WHO.

WHO. 2011. Obesity and overweight. Fact sheet N°311. Updated March 2011. Geneva.

WHO, FAO, UNU. 2007. Protein and amino acid requirements in human nutrition (PDF). WHO Press (available at http://whqlibdoc.who.int/trs/WHO_TRS_935_eng.pdf. Retrieved 2008-07-08).

World Bank, FAO, IFPRI, OIE. 2006. Enhancing Control of Highly Pathogenic Avian Influenza in Developing Countries through Compensation: Issues and Good Practice. Washington: World Bank (available at http://www.fao.org/docs/eims/upload//217132/gui_hpai_compensation.pdf).

World Bank, Latin America & the Caribbean. Undated. What are the facts about rising food prices and their effect on the region? (available at http://go.worldbank.org/CJYWKZPMX)

World Bank. 2003. *From Goats to Coats: Institutional Reform in Mongolia's Cashmere Sector.* Report No. 26240-MOG. Poverty Reduction and Economic Management Unit East Asia and Pacific Region

World Bank. 2009. Mongolia Livestock Sector Study Volume I – Synthesis Report. Sustainable Development Department East Asia and Pacific Region. World Bank: Washington DC.

World Population Prospects. 2002 revision. (available at http://www.un.org/esa/population/publications/wpp2002/WPP2002_VOL_3.pdf).

Yach, D., Stuckler, D. & Brownell, D. 2006. Epidemiologic and economic consequences of the global epidemics of obesity and diabetes *Nature Medicine* - 12, 62 - 66 (2006)

Yi-Zhong, C. & Zhong, Z. Undated. Shanghai: trends towards specialised and Capital-intensive urban agriculture. RUAF city case study Shanghai (available at http://www.ruaf.org/sites/default/files/Shanghai.PDF).

Yi-Zhong, C. & Zhangen Z. 2000. Shanghai: Trends towards specialised and capital-intense urban agriculture. In Growing Cities, Growing Food: Urban Agriculture on the Policy Agenda. N. Bakker, M. Dubbeling, S. Gündel,U. Sabel-Koschella & H. de Zeeuw, (eds) *A Reader on Urban Agriculture*, German Foundation for International Development (DSE), Feldafing, Germany. 467-75.

Yu, S.M., Zhu, L.F., Ouyang, Y.N., Xu, D.H. & Jin, Q.Y. 2008. Effects of rice-duck farming system on biotic populations in paddy field. (article in Chinese) Ying Yong Sheng Tai Xue Bao 2008. Apr;19(4):807-12.